CASE STUDIES IN INDUSTRIAL HYGIENE

CASE STUDIES IN INDUSTRIAL HYGIENE

Edited by

JIMMY L. PERKINS

and

VERNON E. ROSE

University of Alabama at Birmingham
Birmingham, Alabama

A Wiley-Interscience Publication

JOHN WILEY & SONS

New York • Chichester • Brisbane • Toronto • Singapore

Library of Congress Cataloging in Publication Data:

Case studies in industrial hygiene.
"A Wiley-Interscience publication."
Includes bibliographies.
1. Industrial hygiene—Case studies. I. Perkins, Jimmy L. (Jimmy Lee), 1951- II. Rose, Vernon E.
[DNLM: 1. Occupational Medicine—case studies.
WA 400 C337]
RC963.3.C38 1986 613.6′2 86-15709
ISBN 0-471-84263-X

Printed in the United States of America

10 9 8 7 6 5 4 3 2 1

CONTRIBUTORS

JIMMY L. PERKINS, Ph.D., CIH, Associate Professor, University of Alabama at Birmingham, Birmingham, Alabama

VERNON E. ROSE, Dr.PH., CIH, Professor, University of Alabama at Birmingham, Birmingham, Alabama

RAYMOND H. DEIBERT, CIH, Birmingham, Alabama

ARTHUR B. HOLCOMB, Captain, U.S. Army, Redstone Arsenal, Huntsville, Alabama

R. KENT OESTENSTAD, CIH, University of Alabama at Birmingham, Birmingham, Alabama

MAX L. RICHARD, Director of Occupational Health and Safety Program, University of Alabama at Birmingham, Birmingham, Alabama

MICHAEL C. RIDGE, CIH, University of Alabama at Birmingham, Birmingham, Alabama

PREFACE

In 1978, one of us (VR) started an industrial hygiene educational program at the School of Public Health, University of Alabama at Birmingham. Certainly we were not the first to do so, as many other programs were begun before 1978. Necessarily then, we were not the first to discover an important lesson when developing our industrial hygiene curriculum. The essence of that lesson lies in the fact that most industrial hygienists are generalists and are considered "jacks of all trades, masters of none." The industrial hygienist is also the epitome of the applied scientist. He or she applies the principles of toxicology, physiology, physical chemistry, physics, meteorology, organic chemistry, biochemistry, analytical chemistry, and a host of other disciplines to the everyday problems found in the workplace.

How does one go about teaching a student to become an industrial hygienist? Obviously, much of the answer lies in experience, which has been, and will probably continue to be, the greatest "teacher" in our field. The challenge to the educator is to then determine how experience can be translated or better transformed to the formal educational setting. For a student to gain this experience, or even gain the insight of the experience of others, and remain a student, presents somewhat of a dilemma.

Whether we realized the dilemma in 1978 or were simply fortuitous, in 1979 a course was added to the curriculum entitled "Industrial Hygiene Case Studies." It, along with a three-to-six-month field internship, provides a basis for the experience that students will continue to obtain throughout their careers; the experience that makes them industrial hygienists.

Originally the case studies course was designed around NIOSH Health Hazard Evaluations (HHEs). The information in several HHE reports was divided into three sections, which as closely as possible provided the information involved in the three classic industrial hygiene steps of recognition, evaluation, and control. A brief introduction gave some information to the student; it outlined the problem and served as the information that might be developed in a walk-through survey. The second, or evaluation step usually provided air or medical monitoring data developed in response to the information contained in the first section. Finally, the third section presented the recommendations made by NIOSH in terms of controlling existent or potential problems. The students would take each section separately and would come up with their own solutions to the problems, which would then be compared to NIOSH's solutions. While the basic approach worked well, the information provided in the HHE was often not of sufficient depth or detail to provide a thorough learning experience. Often the students had more questions than the HHE had answers. Because of the finite class time available, development of potential answers to the questions was not always possible, thus leaving the students with a sense that the final story had not been told. (While these questions could be answered or discussed in class, the purpose of the case studies course was to promote independent thinking on the part of the student.)

Consequently, the idea for this book came about after two years of experience with the original set of case studies. Included are seven case studies that are based on real problems, all but one of which were experienced by the authors. That case study is developed from a NIOSH HHE. In some cases the experiences have been slightly modified in order to provide a more interesting problem or a problem that would present a greater challenge to the reader.

In addition to providing written experiences for use by students, we have also intended to provide written experiences that can be of great benefit to the practicing industrial hygienist. Although our initial goal in developing this series of case studies was to attack the problem of transposing experience to the classroom setting for industrial hygiene students, a second and equal goal involving the practicing industrial hygienist took shape as the work developed.

For the purposes of this book, it might be convenient to classify industrial hygienists in one of three categories: those who have recently graduated with a degree in industrial hygiene, those without formal training who have ventured into the profession via the "school of hard knocks," and those with several years of experience and who have perhaps attained certification by the American Board of Industrial Hygiene. For those in the first category, the benefit of this book should be similar to that experienced by those currently studying industrial hygiene. The learning process would be somewhat different in that these industrial hygienists may not have the benefit of thoroughly discussing each case study with their peers. Interaction is important in any learning process and is facilitated in the classroom; nevertheless, an independent approach to learning is still better than none.

For industrial hygienists in the second category, experience is obviously their greatest asset. However, many of those industrial hygienists may have operated in a partial vacuum for many years, doing what they thought would work and continuing to do that which did work. While this strategy is productive, again, interaction with others is the quickest way up the learning curve. Someone else may have already solved the problem with which they are confronted. This book then will provide a set of successful solutions to problems. Admittedly, it provides only one set of solutions and certainly any applied science has more than one solution to a problem. We have tried to prepare the book in this spirit by adding an "other solutions" section to each case study. Hopefully, for this group of industrial hygienists, the book will serve to reinforce ideas that have tended to work or to promote alternative solutions for unsolved problems.

Perhaps one of the most important uses of the book is related to those industrial hygienists in category three. Industrial hygiene, being an applied field, is oftentimes plagued with a lack of communication. People tend to publish information concerning theory or design, while information concerning applied knowledge or applied solutions goes on the back burner, never to be published. Consequently, we hope to promote discussion and debate concerning techniques, recipes, methods, and philosophies used by experienced industrial hygienists. Hopefully, this type of effort will flourish, as exemplified by the new journal of *Applied Industrial Hygiene* published by the American Conference of Governmental Industrial Hygienists.

For practicing industrial hygienists, the technique for using the book should be fairly straightforward. The recognition, evaluation, and control sections of each case study should be read with intermittent pauses. If they are read in rapid succession, much of the learning process is lost. Consequently, it is important that after reading the recognition section of a case study, the reader give serious thought to what should be accomplished in the evaluation step, prior to reading the evaluation section. The same careful consideration is then given to the evaluation section before turning to the control section. After reading the final section, the industrial hygienist should give careful consideration to how the entire case study fits together and relates to his or her experience or current needs. This type of approach should not only encourage creative thinking on the part of the reader but should also allow the reader to be critical to the greatest extent possible.

For the student and teacher, the instructions for use of the book are somewhat more involved and depend upon the degree of formality the instructor desires. We have found it useful to have the students prepare detailed reports for presentation in class. The questions contained at the end of each section are considered *only* as starting points for the student. Reports are expected to go beyond these questions and be thorough and professional.

Details were mentioned earlier. By details, we mean that the student should not just point out that a respirator should be used in a given operation but should state the type of respirator; we expect students to state a brand and model. This, we believe, gets them past the theoretical and into the practical things they will have to do on the job. Likewise, the student should not state that "toluene should be sampled for and analyzed," but should state the particulars of the sampling method such as flow rate, specific collection medium, and the particulars of the analytical method, such as gas chromatography using a flame ionization detector. Students must, of course, be able to defend their recommendations. It is important for the classroom setting that the instructor spend some time discussing the overall approach used in the case study versus his own ideas or solutions and those of the students.

With regard to format of the book, we admit some concerns. Formerly, we handed each section of a case study, separate from the other sections, to the class. Therefore, a student looking at the evaluation section of a case study would not have access to the controls section. With the book formatted as it is, students working on the recognition section of a case study have the opportunity to read ahead to the evaluation section in order to find answers for their assignment. While undoubtedly this will occur, its occurrence and effect should be minimized by several factors.

First, most of the students using the text will be graduate students who, hopefully, have matured and recognize the need to learn, rather than get a passing grade in a course. Consequently, to read ahead will have the effect of cheating one's self. Also, this course is designed to be the last course in a sequence of industrial hygiene courses that would constitute the curriculum. By the time this course is to be taken, both the student and the instructor should know whether the student is graduate material and capable of becoming an industrial hygiene professional. It might even be appropriate at this time to share with the student the ethics of our profession (or repeat them if this has already been done) and to impress on the student the need to approach this course as a professional.

Indeed, the essence of the course is to gain experience and challenge one's own thinking by seeking the solutions to problems in an applied science. The bottom line that a student should understand is that failure to use the case studies for their intended purpose will prevent the student from gaining important assets for the furthering of their careers.

JIMMY L. PERKINS
VERNON E. ROSE

Birmingham, Alabama
October 1986

ACKNOWLEDGMENTS

In any effort that requires 3 years to come to fruition, there are obviously many people along the way whose contributions should be acknowledged. Unfortunately, as time progresses, memories become shorter and, consequently, we acknowledge those whom we think provided a significant contribution to this book and express sincere appreciation to all the others without mentioning their names. Perhaps the most important group to provide thoughtful comments and act as guinea pigs were the students of the UAB Industrial Hygiene Program in 1984 and 1985. They read and used as exercises initial drafts of the case studies and in some cases were able to point out unworkable problems. Hopefully, all such errors have been removed, so that the case studies provide challenges that are conquerable.

Each of the case studies was reviewed by at least two people who were knowledgeable in the area of that case study. Obviously, we are responsible for the contents of the book, and the reviewers could only help make the chapters better and in no way make them worse. Any errors contained herein are ours. The following gave thoughtful comments as reviewers of this book: Mr. Loren Anderson, CIH, IBM Corporation; Mr. Jeff Burton, PE, Ive, Inc.; Mr. Harry Ettinger, PE, CIH, Los Alamos National Laboratory; Mr. Steve Henry, CIH, Anniston,

Alabama; Dr. Jack Hinton, Texaco, Inc.; Mr. D.G. Hodgkins, CIH, General Motors Corporation; Dr. Nelson Leidel, National Institute for Occupational Safety and Health; Mr. Lester Levin, CIH, Drexel University; Dr. Frank Mitchell, DO, National Institute for Occupational Safety and Health; Mr. W. Mitchell Sams, Jr., M.D., University of Alabama at Birmingham; Mr. Gene Saywell, CIH, General Electric Company; Mr. H. Keith Thompson, CIH, Caterpillar Tractor Company; Dr. Frederick Toca, CIH, U.S. Steel Corporation; Mr. Thomas J. Walker, CIH, Thomas J. Walker, Inc. Two other reviewers chose to remain anonymous.

Ms. Donna Tracy and Ms. Joan Sung spent many hours in front of the word processor and are responsible for the typing of this document. We wish to thank those named, as well as any whom we have inadvertently failed to mention, for their valuable contributions.

J.L.P.
V.E.R.

CONTENTS

1

MANUFACTURE OF LEADED STEEL

RAYMOND H. DEIBERT

VERNON E. ROSE

1.1. PROBLEM RECOGNITION

Process Description

This problem is set in a typical "integrated" steel mill where steel products are made from the basic raw materials, that is, iron ore, limestone, and coal. Scrap metal and various addition or alloying agents may also be used. At the beginning of the process, coal is heated in slot-type ovens and forms a hard, fused carbon product called coke. Iron ore, limestone, and coke are then charged into the top of a large vertical furnace called a blast furnace, and hot air is blown in at the bottom. The resulting chemical reactions reduce the iron ore to molten iron. Impurities are removed in the molten limestone slag, which is also produced. To make steel, the molten iron is refined to adjust the carbon content and to further remove impurities. In modern mills, the refining is done in a basic oxygen vessel or an electric arc furnace; the open hearth and Bessemer furnace processes are obsolete. The finished molten steel is then either poured into ingot molds or processed through a continuous caster to produce shapes that can be further processed.

In the finishing end of the mill, the cast ingots, slabs, or "blooms" (cast shapes of square cross-section) are placed in furnaces and heated to a temperature that allows the shape to be formed into the desired product. Most products are formed in high-speed rolling mills of various sizes and types, while some products are forged under pressure in hammer or forging mills. Different metallurgical properties may be produced by subsequent heat treating (annealing), quenching, and cold rolling. Finished steel mill products include bars, structural members, plate, sheet, rods, wire, and pipe. Sheet products can also be coated in the mill to produce tin plate and galvanized (zinc-coated) steel.

This problem involves a mill's steel-making shop which, in this case, is a "basic oxygen process" or BOP shop. A typical shop will contain two or three, 200-ton BOP vessels, which can produce a "heat" of molten steel in about 30 min. A charge will consist of iron from the blast furnace, scrap, and limestone or dolomite fluxes. Common addition agents are the ferro-alloys of manganese, silicon, chromium, vanadium, molybdenum, and titanium, and metallic nickel, copper, lead, and aluminum. These may be added to the melt in the vessel, to the ladle after tapping, or to each ingot mold as it is filled. Oxygen is blown into the melt from a lance extended into the BOP vessel from above in order to remove the excess carbon; the resulting exhaust gases can contain carbon monoxide, sulfur dioxide, and various particulates.

Leaded steel is a special steel mill product. The lead content acts as an internal lubricant that reduces tool-to-metal friction and thus increases machine tool life. The lead content also helps to improve the machined surface and to produce well-broken, fine, cutting chips, which help in chip handling and disposal. To further improve the machinability of low-carbon steels, other additives such as sulfur, bismuth, selenium, and tellurium are often added along with lead. The addition is usually made by blowing closely sized, high-purity lead shot into the molten steel as it is poured or "teemed" from the transfer ladle into each individual ingot mold. Three to seven pounds of lead shot are added per ton of steel, and the lead content of the finished steel is in the range of 0.15 to 0.35%.

Nature of the Problem

The problem in this case study involved a union complaint that alleged excessive exposure to lead and other metals in the leaded steel manufacturing area of a large steel company. As a consulting industrial hygienist, you are brought in by the joint labor - management safety and health committee to conduct an independent evaluation of the potential health hazards associated with exposure to the metals. During the walkthrough survey of the steel-making process, you observe the following operating conditions.

1. Leaded steel is teemed only at one pouring stand (see Figure 1.1). No more than three leaded steel heats, each containing an average of 210 tons, are poured during any one shift. The pouring operation involves bringing a ladle filled with molten steel to the pouring stand by means of an overhead crane, which is equipped with an air conditioner with a standard dust filter. The steel is then poured (actually it is dropped, see Figure 1.2) into a se-

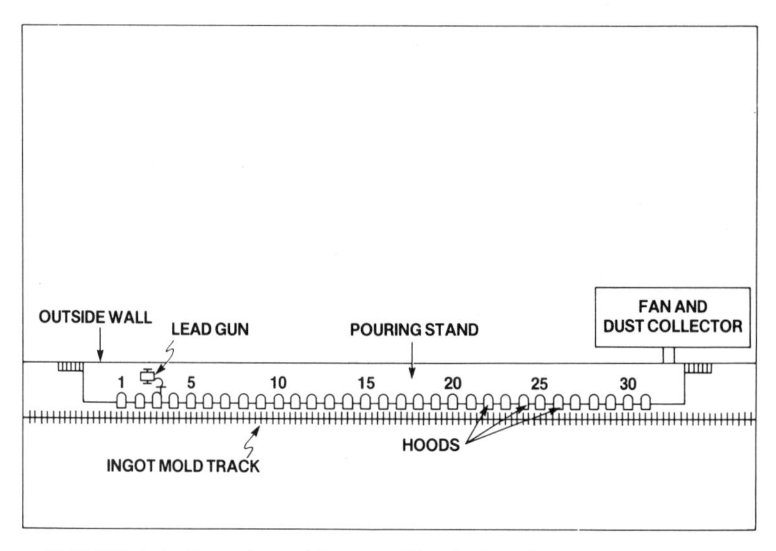

FIGURE 1.1. Top view of layout of leaded steel pouring operation.

ries of cast-iron ingot molds with the concurrent addition of lead shot, using an air-blast gun. The ingot molds are brought into and out of the shop on a railroad track that borders the pouring stand. From 25 to 29 ingot molds are filled equally in each heat.

2. The pouring crew usually consists of six members. Their job titles and assignments are outlined as follows:

Craneman	Works from crane cab; positions the ladle over each mold as it is filled.
Steel pourer	Operates the pneumatic ladle stopper controls; responsible for the correct filling of each mold.
Pitman	There are three pitmen who rotate between the following jobs on a per-shift basis. Lead-gun operator: Makes the lead addi-

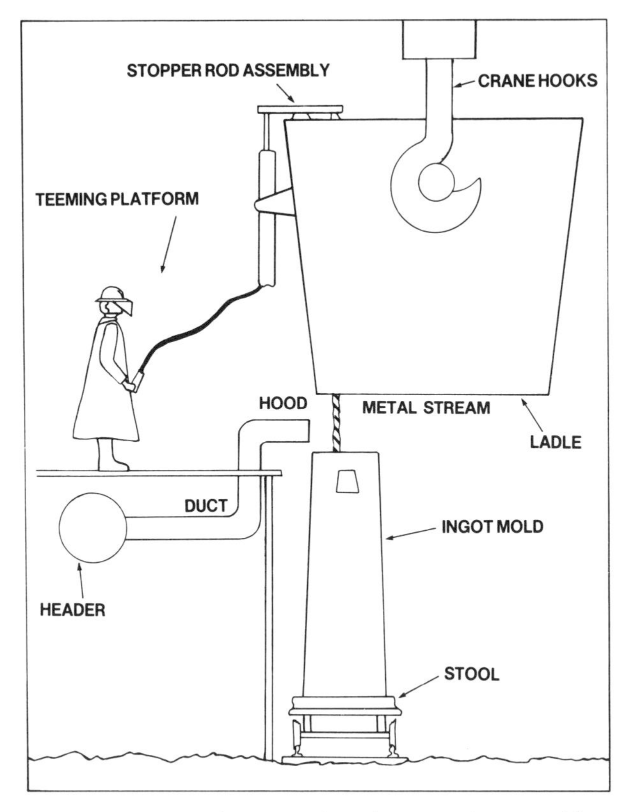

FIGURE 1.2. Schematic of workers teeming a mold.

Pitman (continued) tions to each ingot mold as it is filled with molten steel.

Topping man: Puts a granular insulation material on the top of the steel after the mold is filled; helps fill lead gun.

Sample man: Takes a sample of the steel as the mold is filled; helps fill lead gun; opens and closes the hood dampers on the local exhaust ventilation system.

Metallurgical observer	Identifies and makes a record on each ingot as it is poured.

In addition, the crew foreman and a second metallurgical observer may also be present. With the exception of the craneman, the pouring crew work on the pouring stand near each other, and are usually within 10 ft of the ingot mold as it is being filled.

3. When pouring leaded steel, about 25 to 50 lb of lead shot are added to each ingot when it is one-third to two-thirds full. Lead is blown into the steel stream with a "lead gun," which consists of a lead shot hopper and mechanical ejector connected to a steel nozzle by a short length of rubber tubing. The device is mounted on wheels so that it can be easily moved along the pouring stand. Lead shot is added to the hopper as needed from 50-lb bags stored on the pouring stand. In addition to lead, tellurium metal in 8-lb bags may also be added as another steel improving agent. About 30 to 35 min are needed to pour a heat of 25 to 29 molds.
4. The pouring stand where the lead additions are made has a local exhaust ventilation system. Adjacent to each ingot mold, there is a 26 by 11 in. unflanged lateral exhaust hood. The face of the hood is 18 in. from the near edge of the mold. Each of the 31 hoods is designed to exhaust approximately 6000 cubic feet of air per minute (cfm). Since allowing each hood to be open during the entire teeming operation would be inefficient, they are opened individually according to a prescribed sequence. When a particular mold is being filled, the exhaust hoods for that mold, the two preceding ones, and the one about to be filled are opened by operating the hood dampers. The sample man operates the hood dampers. Each hood exhausts mold emissions for about 4 min and

is then blocked off. Fumes captured by the hoods are conveyed to nearby collecting hoppers. The recovered solids are shaken into bins lined with polyethylene bags. The bins are removed when full by other employees who are required to wear respirators. No determination could be made concerning the exact nature of the respirator. The pitmen sweep down the pouring stand once per week on the daytime shift.

5. The following observations were also made by the industrial hygienist during the initial survey:

Candy bar wrappers and cigarette butts were seen on and around the pouring stand.

During one heat, tellurium metal in 8-lb bags was also added as another steel improving agent.

Because of the radiant heat load during the pouring operation, the crew on the pouring stand wore aluminized coats, leggings, gloves, and hoods with face shields (see Figure 1.2). No respirator usage was observed on the pouring stand.

A shower of sparks was thrown up when the steel pourer accidentally allowed the steel stream from the ladle to hit the top rim of an ingot mold.

The shop building was very large and open on the inside. Air could enter from many openings, including those provided for railroad and vehicle entry.

A red- or orange-colored fume evolved when each ingot mold was filled. Some of the fume cloud seemed to escape capture by the ventilation system.

The crew appeared to be diligent in operating the ventilation system dampers as required.

The crew lunchroom and rest area was located in a utility building about 50 ft from the end of the pouring platform.

The room appeared to be reasonably clean and sanitary. There were no shower facilities available.

Questions

1. For what potential air contaminants will you sample?
2. What sampling equipment will you use? Be specific as to type and quantity.
3. How will you sample: short term or full shift; personal or area? Will the crew's aluminized clothing pose a problem to sampling? Will metal sparks or radiant heat pose a problem?
4. How will you test the performance of the ventilation system? What equipment will you use?
5. Are medical records and biological monitoring of any help to you? What test results would you like to see?

1.2. EVALUATION RESULTS

Preliminary Findings

The plant medical director was visited to determine if a biological monitoring program had been implemented for the workers involved in the leaded steel operation. A blood lead monitoring program for the crew involved in pouring leaded steel had recently been implemented; the most recent biological monitoring results for the involved workers are shown in Table 1.1.

In-Depth Studies

Based on the results of the preliminary survey and the results of the biological monitoring program, the industrial hygienist determined that it would be appropriate to conduct an in-depth study involving air monitoring for exposures to lead, iron oxide, and tellurium fumes. It was also determined that a thorough evaluation of the ventilation system would be appropriate.

Air samples were collected with personal sampling pumps operated at a flow rate of 2.0 L/min over at least a 7-hr period. This sampling procedure is based on that recommended by OSHA. Twelve pumps were calibrated for use as personal, area, and some spare samplers. Pump airflow was calibrated prior to and following the survey using a bubble meter.

The collection medium employed was a 37-mm diameter Millipore Type AA filter (mixed cellulose ester), with an effective pore size of 0.8 μm. The small pore size was selected because the particulates to be encountered would be metal fumes. Fiberglass filters, efficient in collecting 0.3-μm spheres, could have been used as well. The filters, with backup pads, were packed in two-piece field-monitor cassettes.

The survey was conducted over three consecutive daytime shifts. Full-shift, personal monitor samples provided the primary data for the exposure evaluation. The personal monitor samples were from wearer's breathing zone, that is, from within 1 ft of the wearer's mouth and nose with no in-between obstructions. To accomplish this, the crew members had to keep the filter cassette inside their aluminized hood face shield. The industrial hygienist instructed them to do this and was present in the event that they needed help. Keeping the cassette inside the face shield also protected it from radiant heat and metal sparks. Pump flow rate was checked using the integral rotameter each hour and adjusted as necessary.

Because dry sweeping of toxic dusts may significantly contribute to exposure levels, separate measurements for exposure levels during sweeping of the pouring stand floor were determined to be important. Plans were made to collect 30-min, short-term personal monitor samples to measure the lead ex-

TABLE 1.1
Results of Biological Analyses (Averages of Last Three Measurements)

Subject	Blood Lead Level (μg/100 mL rbc)	Hemoglobin Level (g/100 mL)	Hematocrit (vol. %)	Free Erythrocyte Protoporphyrin (μg/100 mL rbc)	BUN (mg/100 mL)	Creatinine Level (mg/100 mL)
Metallurgist	30	16.9	50.1	22	12	1.3
Craneman	24	14.6	45.8	163	16	1.2
Gun operator	63	14.8	45.6	123	15	1.1
Topper	42	17.3	52.4	31	17	1.0
Gun operator	60	13.9	43.6	358	14	1.2
Topper	54	15.4	46.4	189	13	1.0
Foreman	27	16.2	48.7	45	20	1.5
Sampler	88	16.3	48.5	142	16	1.4
Metallurgist	30	17.5	52.5	14	15	1.1
Craneman	20	14.4	43.3	29	10	0.9
Foreman	33	15.2	46.0	74	16	1.2
Pourer	35	14.4	43.7	115	12	1.4
Pourer	46	15.7	48.5	91	12	1.1
Foreman	41	15.1	45.5	56	16	1.2
Craneman	18	16.0	48.1	19	16	1.3
Sampler	82	15.5	47.7	148	18	1.4

posure of the pitmen when they were sweeping the pouring-stand floor. The analysis method had sufficient sensitivity to permit short-term sampling. For this brief time, the pitmen would be required to wear two sampling pumps. On the third day of the survey, it was learned that a pair of laborers would empty the dust-collection bins on the pouring-stand ventilation system. Arrangements were made to collect short-term, personal samples during this operation. During this part of the survey, the industrial hygienist noted that the laborers who cleaned the dust bins wore disposable nuisance-dust respirators.

Area samples are usually of limited value, but they can provide data to corroborate the personal monitoring data and aid in determining contamination sources. Therefore, three samplers were stationed at points along the pouring stand, and one was attached to the lead-gun hopper. The filter cassettes were mounted at breathing height. Area samples were also collected in the lunchroom. All air-sampling results are shown in Table 1.2.

Testing the ventilation system was a necessity, since the efficiency of the system would have great bearing on the magnitude of the lead exposures. Before the survey, a review was made of the section on testing of ventilation systems in the *Industrial Ventilation Manual*, published by the American Conference of Governmental Industrial Hygienists. A Pitot traverse inside the ductwork is recommended as the best test method, but this method was judged to be impractical because of the bends in the system ductwork. Instead, a velocity traverse across each hood face was made. The air volume (cfm) exhausted could then be determined from the average velocity measured (ft/min) and the area of the hood opening (ft^2). For a duct of this size, ACGIH recommends that at least 16 velocity readings be taken at the center points of equal areas across the hood face. Velocity readings were made with a Sierra Air Velocity Meter, Model 441. The meter had been recently cali-

TABLE 1.2
Air Sampling Results

Date	Job Title or Area Description	Sample Duration (min)	Lead ($\mu g/m^3$)	Tellurium (mg/m)[a]	Iron Oxide Fume (mg/m)[b]
12/6	Pit foreman	420	150	0.021	0.8
12/6	Craneman	482	40	0.012	0.2
12/6	Metallurgist	422	25	0.007	0.9
12/6	Steel pourer	440	85	0.019	1.2
12/6	Pitman: gun operator	422	220	0.003	2.0
12/6	Pitman: topping man	430	14,000	0.014	2.9
12/6	Pitman: sampler	428	670	0.007	4.1
12/6	In lunchroom, on table	441	12	0.004	0.3
12/6	On lead-gun unit	430	850	0.009	1.0
12/6	North stand, across from 6th hood	444	190	0.024	1.6
12/6	North stand, across from 16th hood	446	350	0.040	3.0
12/6	North stand, across from 28th hood	448	70	0.022	1.4
12/7	Pit foreman	431	45	—	0.7
12/7	Craneman	485	9	—	0.6
12/7	Metallurgist	420	18	—	0.9
12/7	Steel pourer	431	41	—	2.3
12/7	Pitman: gun operator	425	93	—	1.8
12/7	Pitman: topping man	420	73	—	2.5
12/7	Pitman: sampler	429	38	—	3.1

12/7	In lunchroom, on table	432	7	—	0.1
12/7	On lead gun unit	426	210	—	1.5
12/7	North stand, across from 6th hood	429	85	—	1.9
12/7	North stand, across from 16th hood	437	130	—	2.8
12/7	North stand, across from 28th hood	421	19	—	2.1
12/8	Pit foreman	422	31	—	0.6
12/8	Craneman	484	15	—	0.3
12/8	Metallurgist	431	45	—	0.7
12/8	Steel pourer	428	72	—	2.7
12/8	Pitman: gun operator	430	140	—	3.1
12/8	Pitman: topping man	425	95	—	2.4
12/8	Pitman: sampler	423	160	—	1.6
12/8	Laborer, while changing ventilation system collection bins	61	425	—	(17.9)[b]
12/8	Laborer, as above	64	460	—	(23.0)[b]
12/8	Pitman, while sweeping off north stand platform; job done on daytime shift once per week	44	55	—	(1.3)[b]
12/8	Pitman, as above	46	37	—	(0.5)[b]
12/8	Pitman, as above	45	30	—	(1.0)[b]

[a]— Signifies that tellurium was not used on these days.
[b]Probably iron oxide dust and not iron oxide fume.

brated against a Kurz Air Velocity Calibration System (Model 400A). The results of the ventilation measurements are shown in Table 1.3.

While evaluating the ventilation systems, the industrial hygienist noted two occasions where the train crew spotted the ingot molds off center from the ventilation system hoods. Also, the maintenance foreman could not remember the last time the ventilation system ductwork had been inspected or cleaned.

Questions

All the survey data and other information is now in your hands. Analyze the data and other information, draw your conclusions, and make necessary recommendations for control. The following specific questions should guide your thinking.

1. Are any members of the pouring crew exposed to excessive levels of lead, tellurium, or iron oxide fume? What data do you have to support these conclusions? Should any of these workers be removed from further exposure to lead?
2. Is the ventilation system operating properly?
3. Finally, what are your recommendations for additional control?
   a. In the area of engineering control, what should be done about the ventilation system?
   b. Can certain work practices be changed to improve the situation?
   c. Do you have any recommendations regarding personal protection?
   d. Should industrial hygiene and medical monitoring be continued?

TABLE 1.3
Ventilation Measurements: Exhaust Hoods, North Pouring Stand

Hood No.[a]	Average Measured Velocity[b] (ft/min)
1	750
2	775
3	650
4	1100
5	900
6	725
7	1400
8	1000
9	1025
10	1050
11	1375
12	1000
13	1275
14	1250
15	1075
16	1375
17	1250
18	1375
19	1550
20	1625
21	1700
22	1750
23	1700
24	1850
25	1800
26	1875
27	1875
28	1825
29	2000
30	2025
31	2050

[a]Hood face dimensions are 26 by 11 in.

[b]Face velocities at each hood (except 1, 30, and 31) were measured with the dampers of the preceding and two following hoods open and all other dampers closed. Face velocities on hood 1 were measured with hoods 2 and 3 in operation. Face velocities for hood 30 were measured with hoods 29 and 31 in operation. For hood 31, hoods 29 and 30 were also in operation.

1.3. CONTROL STRATEGIES

Discussion

Airborne contaminant levels were first compared with the following OSHA standards.

Lead (inorganic)	50 $\mu g/m^3$ (micrograms per cubic meter) as an 8-hr, time-weighted average permissible limit. 30 $\mu g/m^3$ as an 8-hr, time-weighted average "action level," which triggers certain precautionary measures.
Tellurium	0.1 mg/m^3 (milligrams per cubic meter) as an 8-hr, time-weighted average permissible limit.
Iron oxide fume	10 mg/m^3 as an 8-hr, time-weighted average permissible limit (the current ACGIH TLV for iron oxide fume is 5 mg/m^3 as an 8-hr TWA).

Personal sample results indicate the following pouring crew occupations were exposed to lead levels above the OSHA 50 $\mu g/m^3$ permissible limit on one or more of the three shifts sampled.

Pit foreman (1 shift)

Steel pourer (2 shifts)

Pitman: Gun operator (3 shifts)

Pitman: Topping man (2 shifts*)

Pitman: Sampler (2 shifts)

*A third sample indicated an exposure level of 14,000 $\mu g/m^3$. This enormously high result, far out of line with the others, is likely due to laboratory error or sample contamination.

The crane operator and the metallurgist were not exposed to lead levels above the 50 $\mu g/m^3$ permissible limit. However, both occupations were exposed to levels above the 30 $\mu g/m^3$ action level during one of the three shifts sampled. Personal samples collected on the laborers while changing the pouring-stand ventilation system collection bins indicated an excessive exposure is produced by this operation alone, even though it lasts for only about 1 hr per shift. Personal samples collected on the two pitmen while they swept the pouring platform indicated a measurable, but minor, exposure to lead during this 1-hr-per-week job.

Results from area samplers stationed on the lead gun and along the pouring platform indicated very high lead levels throughout the area. Lead levels appeared to increase with the distance from the ventilation system fan. Lead levels inside the crew lunchroom were well below both the OSHA permissible limit and action level, suggesting that time spent in this area would not significantly contribute to the worker's overall exposure level. This assumption, however, is based only on air sampling results. The industrial hygienist must also be aware of ingestion as a source of lead uptake. Washing of hands before eating, cleanliness of the lunchroom and break areas, and prohibitions of eating or smoking in the work area are important in reducing lead uptake from ingestion.

Recent blood lead monitoring of crew members indicated that 8 of 16 crew members have elevated blood lead levels (above 40 $\mu g/100$ mL whole blood). These findings are consistent with the air sampling data in that those with excessive blood lead levels were from personnel having the greatest exposure to airborne levels of lead. Under the OSHA lead standard (29 CFR 1910.1025) the five employees with blood lead levels in excess of 50 $\mu g/100$ g should be removed from further occupational exposure to lead (exposure levels less than 30 $\mu g/m^3$ for an 8-hr TWA) and should not return to their jobs until their blood lead levels are below 40 $\mu g/100$ g. In addition, their exposure levels to airborne lead must not exceed the OSHA PEL. For the five workers with blood lead levels between 40 and 60, blood leads must be repeated every 2 mo.

The pouring crew members were exposed to tellurium levels and to iron oxide fume levels well within OSHA (and ACGIH) permissible limits during the shifts sampled. The laborers changing out the ventilation system collection bins were exposed to relatively high iron oxide dust levels, but only for 1 hr/day. On an 8-hr, time-weighted average basis, their exposure to iron oxide dust was not excessive.

Ventilation system measurements indicated that the ingot mold hoods are exhausting from 1300 to 4100 cfm, with the exhaust rate being much lower at the end of the system farthest from the exhaust fan. These measured rates are all lower than the 6000 cfm design rate.

From the above results it can be concluded that the pouring crew is exposed to excessive lead levels, which is corroborated by the biological monitoring results. It appears that the excessive exposures are due to a combination of factors, including a ventilation system that is operating far below design capacity, as well as poor work practices. It was also noted that where personal protective equipment was used to reduce exposure levels to airborne contaminants, improper equipment had been provided.

Ventilation system performance should be brought up to at least design capacity, that is, 6000 cfm of air flow through each hood. Survey results suggest leakage and/or blockage to be a problem. System clean-out and repair may be all that is needed. The fan specifications and blade should be checked to ensure that the system installed meets design specifications.

Short-Term Controls

The industrial hygienist recommended that three immediate steps be taken with a goal of reducing worker exposure levels to airborne lead fumes. These measures involved changes in work practices and the use of appropriate personal protective equipment, including respirators and clothing.

The most immediate work practice change to be implemented should be to assure that the ingot mold train crews locate the ingot molds directly in front of the ventilation hoods. During the survey, the molds were occasionally spotted out of position, thus greatly reducing the capture efficiency of the ventilation system.

The second work practice of concern involved eating, drinking, smoking, and the use of cosmetics on and around the pouring platform. These acts should be prohibited since they can result in the intake of lead through ingestion.

Finally, it was thought important to immediately initiate a comprehensive respirator control program for those workers whose exposure to airborne lead fumes exceed the OSHA Permissible Exposure Limit of 50 $\mu g/m^3$. All of the pouring crew, with the exception of the crane operator and metallurgist, should be involved in the program. The latter two need not be required to wear respirators, but their voluntary participation in the program should be discussed with them. As a minimum, these programs must comply with the specific regulatory requirements as contained in the OSHA standard on lead exposures (1910.1025 (f)) and respiratory protection (29 CFR 1910.134 (b), (d), (e), and (f)). Since the lead exposure levels are not in excess of 500 $\mu g/m^3$, half-mask, air-purifying respirators equipped with high-efficiency particulate filters can be used. Respirators must be selected from those approved for protection against lead dust, fume, and mist by the Mine Safety and Health Administration and the National Institute for Occupational Safety and Health. The OSHA standard also requires that where workers are exposed to lead above the PEL, it is necessary to provide them with protective work clothing. The clothing in use for protection against radiant heat will satisfy this requirement.

In addition to these immediate steps involving work practices and use of respirators, the employer should thoroughly review the OSHA Lead Standard to determine what other steps are required to be implemented in this workplace. Of specific concern is the need to notify employees of their exposure monitoring results within 5 working days from the receipt of those results. The OSHA standard requires that such notification be in writing, and that those workers whose exposure was in excess of the permissible exposure level must also be advised as to the description of the corrective action to be taken to reduce exposure to below the permissible exposure limit. The employer must also develop a compliance program with both short- and long-term goals and maintain this written compliance program in the workplace. The other requirements for change rooms, showers, lunchrooms, and laboratories as contained in Section (i) of the OSHA lead standard also need to be reviewed and, as appropriate, implemented in this workplace. Of great importance is the need to design and implement an employee education program to assure that workers are aware of the health hazards associated with their jobs, the controls management has and will implement to reduce exposures, and what they themselves can do to limit their exposures.

Long-Term Controls

In addition to the short-term control measures described above, the employer should give consideration to the other requirements contained in the OSHA Lead Standard. Of specific concern is the need to assure that medical examination of workers exposed above the Action Level continues to be in place, and where excessive blood levels are encountered, that a medical removal program is instituted.

Of most vital concern for longer-term controls is the need to evaluate the problems associated with the ventilation system.

It is possible that the low volume of air-flow problems are due to inadequate maintenance, damage to the system, or other mechanical problems, and that the system would operate at design levels if corrective action were taken. It may also be necessary to consider resizing of certain ducts, the installation of flanges on the hoods (which might increase face velocities by up to 10%), and/or a reevaluation of the adequacy of the fan. In addition to these measures, the employer may find it necessary to reevaluate the need for showers, change rooms, and other permanent facilities required in the OSHA standard.

As a final step, the industrial hygienist should also investigate potential lead exposures at subsequent rolling and finishing operations, that is, bar mill, grinding, and scarfing operations.

REFERENCES

American Conference of Governmental Industrial Hygienists. *Industrial Ventilation*, 18th ed., ACGIH, Cincinnati, OH, 1984.

American Conference of Governmental Industrial Hygienists. *Threshold Limit Values for Chemical Substances and Physical Agents in the Work Environment and Biological Exposure Indices with Intended Changes for 1984-1985*. ACGIH, Cincinnati, OH, 1984-1985.

Amstead, B. H., Ostwald, P. F., and Bergeman, M. L. "Production of ferrous metals," in *Manufacturing Processes*, 7th ed., Wiley, New York, 1977.

National Institute for Occupational Safety and Health. *Manual of Analytical Methods*, 2nd ed. Vol. 7. DHEW (NIOSH) Publication No. 82-100, Cincinnati, OH, 1981.

Rulf, R. C. Lead exposure control in the production of leaded steel. *Am. Ind. Hyg. Assoc. J.* 24:63 - 67, 1963.

2

MANUFACTURE OF SUITCASES

VERNON E. ROSE

2.1. PROBLEM RECOGNITION

Process Description

The manufacture of plastic suitcases involves a variety of operations in addition to the initial molding of the suitcase body. In this case study, the manufacturing department receives the molded suitcase bodies, and the following operations are carried out to produce a final product.

1. Hardware casting. Metal suitcase frames are die-cast with a magnesium–aluminum alloy. Molten metal is injected into the die, cooled, and removed. The melting pot, containing the molten alloy and the fluxes, is located next to the casting unit.

2. Putty and file. Defects in cast-metal suitcase frames are filled with a two-part, high-temperature-cured resin. After the resin is cured, the metal and resin are filed and hand sanded. Local exhaust ventilation is not available, but several times during a shift, dust is swept up.
3. Powdered paint operation. The metal frames for certain suitcase designs are painted. The painting is conducted in an enclosed room that contains a ventilated booth. The parts pass through the spray booth on a conveyor. The painter sprays the frames with a dry powdered paint as they pass by.
4. Bright-dip area. The metal suitcase frames are bright-dipped. They are attached to an overhead conveyor and automatically pass through solutions of chromic acid, phosphoric acid, and nitric acid.
5. Plating. Small unassembled stamped metal parts are plated with nickel and chrome (decorative chrome plating), using separate plating baths. The process involves cleaning, plating, and the application of a rust inhibitor.
6. Hardware subassembly. Finished parts, such as locks and catches, are assembled from small parts that may have been top-plated with either chrome or nickel.
7. Buffing. Stamped metal parts, which have been chrome or nickel plated, are buffed using a polishing rouge and a buffing wheel.
8. Welding and hardware maintenance. Welding is conducted in the shop and throughout the factory. Various metals are welded and brazed, including mild steel, stainless steel, and galvanized steel. Silver soldering may also be done.
9. Final repair. As the completed cases reach the end of the assembly line, they are inspected, minor repairs made, and final cleanup provided. The final cleanup primarily

involves the use of a solvent to remove black scuff marks caused by contact with conveyor belts.

Nature of the Problem

The suitcase manufacturing company has decided to change its workers' compensation insurance carrier. As the insurance company industrial hygienist, you make an initial occupational health survey of the plant to determine what potential industrial hygiene problems exist and what actions the company might take to alleviate any problem conditions found. Because of scheduling problems, your initial trip will be limited to one or two days at the plant. Your initial step involved a walk-through survey of the manufacturing process, during which the information involving exposure to airborne contaminants noted in Table 2.1 was determined. While you also note potential problems regarding exposures to heat stress, noise, and other physical agents, these are to be considered at a later date.

In your initial walk-through of the manufacturing area, you are advised by plant management that the workers in several of the departments have noted dermatitis problems. A number of workers have been evaluated as to skin problems, and the findings are available in the medical department.

Questions

1. Assuming that there does appear to be some dermatologic problems among workers in the plant, discuss what approach you might use to further define the scope of the dermatitis problem.
2. What industrial hygiene evaluation techniques would you use to further define the potential health hazards in the nine work areas you surveyed?

TABLE 2.1
Findings of Walk-Through Survey

Work Area	No. of Employees	Potential Health Hazards
1. Hardware casting	2	Aluminum, magnesium/potassium chloride and fluoride, and total particulates
2. Putty and file	32	Skin contact with epoxy resins
3. Powdered paint operation	1	Powered paints which are a mixture of titanium dioxide, calcium carbonate, and polyester resin
4. Bright dip	1	Nitrogen dioxide, chromic acid, and phosphoric acid
5. Plating	2	Nickel and chrome plating solutions, primarily nickel sulfate and concentrated chromic acid
6. Hardware subassembly	100	Skin contact with nickel or chrome-plated parts
7. Buffing	5	Skin contact with nickel- or chrome-plated parts and polishing rouge
8. Maintenance	2	Welding fumes including iron, zinc, cadmium, nickel, and copper
9. Final repair	1	Toluene and 1,1,1-trichloroethane

2.2. EVALUATION RESULTS

Medical Findings

As a result of the findings of the initial survey, the medical investigation focused on dermatologic problems among employees in the putty and file department and the subassembly department. Also included, in a less formal manner, were dermatologic and other medical problems among employees in the buffing, plating, and paint departments, the hardware casting, and the tool and die shop. Private interviews were conducted with employees, including systematic samples of employees in the putty and file department and subassembly department.

The sample of subassembly employees was obtained by choosing, from a seniority list of employees present on the day of the interviews, a random starting position among the first five employees on the list and selecting this person and every fifth person thereafter (without recycling through the list). The sample of putty and file employees was obtained by choosing from an alphabetical list of employees present on the day of the study a random starting position from among the first three employees on the list and selecting this person and every third person thereafter (without recycling through the list). (A seniority list was used in one department, and an alphabetical list in the other, because these happened to be the employee lists available.) The samples were intended to be of sufficient size to allow for a qualitative assessment of the prevalence of dermatologic problems; they were not designed to provide a quantitative assessment in the statistical sense. All 12 subassembly employees in the selected sample were women; 11 were white, 1 was black. Age and seniority data are presented in Table 2.2. Five of the six putty and file employees in the selected sample were women; all six were white. Age and seniority data are presented in Table 2.3.

TABLE 2.2
Employee Interviews in Subassembly Department

Employees in department more than 1 wk		70
No. absent		12
Leave of absence	4	
Vacation	4	
Absent for unknown reasons	4	
Present on day of study		58
Sample		
No. selected (percentage of employees present)		12 (21)
No. interviewed (percentage of sample)		12 (100)

Interviewed Employees	Rash	No Rash	Total
No.	6	6	12
Age (yr)			
Range	26–54	20–57	20–57
Median	38	32	32
Time in department			
Range	2 mo–16 yr	3 wk–24 yr	3 wk–24 yr
Median	3 yr	2 yr	2.75 yr
Time at company			
Range	2.5 mo–16 yr	3 wk–27 yr	3 wk–27.5 yr
Median	7.5 yr	10.5 yr	9.5 yr

Six (50%) of the 12 subassembly employees in the selected sample reported a rash, as did four (67%) of the putty and file employees in the selected sample. In the subassembly department, the employees who had a rash were comparable to those who did not, with respect to age and time at the company, but tended to have been in the department longer. In contrast, the putty and file employees with and without a rash tended to have comparable departmental and company seniority, but those who had a rash were younger than those who did not.

Four other putty and file employees asked to be included in the interview. All were women, and three reported a rash simi-

TABLE 2.3
Employee Interviews in Putty and File Department

Employees in department less than 1 wk	Information not available
No. absent	Information not available
Present on day of study	23
Sample	
No. selected (percentage of employees present)	6 (26)
No. interviewed (percentage of sample)	6 (100)

Interviewed Employees	Rash	No Rash	Total
No.	4	2	6
Age (yr)			
Range	20–31	44–51	20–51
Median	25	48	30
Time in department			
Range	3 wk–1.5 yr	3 wk–7 mo	3 wk–1.5 yr
Median	4.5 mo	4 mo	4.25 mo
Time at company			
Range	4 mo–1.5 yr	6 mo–11 yr	4 mo–11 yr
Median	7 mo	6 yr	9 mo

lar to that of the employees in the selected sample. Two of these also reported nasal congestion, which they associated with their work. Only one of the three, an employee who had recently transferred out of the department, had any observable dermatologic abnormalities: several solitary excoriations (sores caused by, or further damaged by, scratching or picking) and scars on her arms.

In the subassembly department the descriptions of the rash were less consistent than in the putty and file department. In four of six subassembly cases from the selected sample there was erythema, but in only two of these were there papules. Five of the six cases involved arms and/or wrists; hands were involved in two of the five. In the sixth case, the hands, but not the arms, were involved. In this case, and in one other, the face or neck was also involved. There was itching in five of the six cases, and in the remaining one there was a burning sensation. The rash developed 3 wk to 3 yr after the subject began work in the department. Information on the time required for the rash to clear up was inadvertently not obtained. Based on information from three affected employees, the rash recurs within 1 day after returning to work. No consistent relationship between occurrence of the rash and either hot weather or a specific subassembly operation was apparent. Excluding a black discoloration of the hands that results from handling small metal parts, there were dermatologic abnormalities in two of the six affected employees: one had erythema of one wrist, and the other had a few red macules (flat lesions that can be seen but not felt) on one wrist.

Five other subassembly employees asked to be interviewed; four had a rash, and the other had eye and nasal irritation. These rashes were similar to those of the employees in the selected sample, but in three of the four cases the onset did not occur until the employee had been in the department at least 10 yr. One of these five employees, but none of the subassembly employees in the selected sample, reported a dermatologic reaction to jewelry. This reaction consisted of erythema only and was similar to the rash she experienced at work, except that the work-associated rash itched, whereas the jewelry-associated rash did not. As a result of the interview with the subassembly employees, several of the plated parts were obtained, rinsed with water, and the water analyzed for chromium and nickel. Small amounts of both metals were found in the rinse water.

In the putty and file department employees tended to describe their rashes similarly. All four cases from the selected sample consisted of erythema (redness) and papules (small, solid bumps) on the arms, without involvement of the hands; in three cases, the face or neck was involved at least some of the time. In three of the four cases, itching usually accompanied the rash. The rash developed 5 wk to 3 mo after being in the department, cleared up after 2 days to 1 wk of absence, and recurred within 3 or 4 days after returning. There was general agreement that hot weather aggravated the problem, but no consensus on whether long sleeves had a protective effect. None of the sampled employees had a rash on the day of the interviews.

Two buffing department employees were also interviewed. Both were women in their 40s, and both worked with polishing rouge. One reported erythematous macules on the same arm she uses to hold the rouge. The rash, which began after she worked at the job for a year, did not affect her hand. She noted that she wore a glove when holding the rouge. The other worker reported erythematous, coalescent papules on both hands and arms, the rash occurring intermittently over several years. Neither had any dermatologic abnormalities at the time of the interview.

Nine other employees, in four other departments, were interviewed; seven had one or more of a variety of medical problems, one of which was primarily dermatologic but dissimilar to those already described. Four of the seven had obvious potential occupational exposures to various substances in the foundry, plating, or paint departments that might account for their symptoms. The other two employees who worked in the tool and die shop had redness and irritation of their eyes. The cause of their problem was not apparent, although vapors escaping from the paint room were a speculative explanation.

Air Monitoring

The environmental air sampling was designed to determine the workers' 8-hr, time-weighted average exposure to the various airborne contaminants identified during the walk-through survey. In most sampling situations personal monitoring devices were used. In some instances, area sampling at fixed locations was necessary. The sampling and analytical techniques used are presented in Table 2.4, and the results of the air monitoring are shown in Table 2.5.

Questions

1. What are your interpretations of the medical interviews and examinations?
2. What are the applicable evaluation criteria for the air monitoring data? What are your interpretations of the results especially regarding the accuracy of your findings?
3. Based on the medical and environmental monitoring results, what controls, short- and long-term, might you recommend?

2.3. CONTROL STRATEGIES

Discussion

MEDICAL

To the extent that the selected samples of employees represented their respective departments, it would appear that at least half of the subassembly and the putty and file employees had a rash in the recent past. Although the incidence or prevalence of rash in the population from which the employees come is not known, it would not likely be high enough to account for the high rate of reported rashes in these departments, especially since many of the cases seem to have similar features, more notably in the putty and file department.

Neither the medical nor temporal characteristics of the subassembly employees' rashes, nor their work activities or environment, suggest a likely explanation of the rashes. Nickel

TABLE 2.4
Air-Sampling and Analysis Methods

Substance	Collection Media	Flow Rate (Lpm)	Analysis	Reference[a]
Aluminum	Filter[b]	1.5	Atomic absorption	173
Cadmium	Filter[b]	1.5	Atomic absorption	173
Chlorides	Impinger (Acetate buffer solution)	1.0	Selective ion electrode	115
Chromic acid (as chromium trioxide)	Filter[c]	1.5	Colorimetry	S317
Copper fume	Filter[b]	2.0	Atomic absorption	S354
Fluorides	Treated filter	1.5	Selective ion electrode	212
Iron oxide	Filter[b]	1.5	Atomic absorption	S366
Magnesium oxide	Filter[b]	1.5	Atomic absorption	S369
Nickel	Filter[b]	1.5	Atomic absorption	S206
Nitrogen dioxide	Triethanolamine-coated molecular sieve	0.05	Colorimetry	S320
Nuisance particles	Filter[c]	1.5	Total weight	0600[d]
Phosphoric acid	Filter[b]	1.5	Colorimetry	S333
Powdered paint	Filter[c]	1.5	Total weight	0600[d]
Toluene	Charcoal tube (150 mg)	0.20	Gas chromatography	S343
1,1,1-Trichloroethane	Charcoal tube (150 mg)	0.20	Gas chromatography	S328
Welding fume	Filter[c]	1.5	Total weight (see also individual components)	7200[d]
Zinc oxide	Filter[b]	1.5	X-ray diffraction	S316

[a] *NIOSH Manual of Analytical Methods*, 2nd Ed., Vol. 1–7.
[b] Cellulose ester membrane filter.
[c] Poly(vinyl chloride) filter.
[d] *NIOSH Manual of Analytical Methods*, 3rd Ed., Vol. 1 and 2.

TABLE 2.5
Air-Sampling Results

Department	Worker/ Location[a]	Contaminant	Concentration[b]
Casting	Casting machine operator	Aluminum	0.5 mg/m^3
		Magnesium	2.0 mg/m^3
		Total chlorides	0.5 mg/m^3
		Total fluorides	2.0 mg/m^3
		Total particulates	7.0 mg/m^3
Powdered paint	Painter	Total particulates	13 mg/m^3
Bright dip	Operator	Nitrogen dioxide	2.0 ppm[c]
		Nitrogen dioxide	4.1 ppm
		Chromic acid	1.0 μg/m^3
		Phosphoric acid	0.1 mg/m^3
Plating	Operator	Nickel	20 μg/m^3
		Chromic acid	70 μg/m^3
Maintenance	Welder[d]	Cadmium	60 μg/m^3
		Copper	10 μg/m^3
		Iron	3.5 mg/m^3
		Nickel	5.0 μg/m^3
		Zinc	9.5 mg/m^3
Final repair	Clean up	Toluene	2 ppm
		1,1,1-Trichloroethane	400 ppm[c]
		1,1,1-Trichloroethane	110 ppm

[a]Worker with greatest potential exposure selected for evaluation.
[b]8-hr, TWA personal sample unless otherwise noted.
[c]15-min STEL measurement.
[d]Sampler placed inside welding helmet.

dermatitis was considered because employees handle metal parts containing nickel. However, since only one employee—who was not among the selected sample—reported a dermatologic reaction to jewelry, nickel is probably not the cause of the general problem. Allergic chromate dermatitis does not occur if contact with chromium is limited to metallic chrome or chrome alloys, but it could possibly result from residual soluble chromates on inadequately rinsed parts. However, the reported appearance of the rash and its distribution, frequently sparing the unprotected hands, is not suggestive of the irritant effects of chromates. Substances on the surface of the parts received in the subassembly department could conceivably cause dermatitis, but the data from the employees are not particularly suggestive of either irritant or allergic contact dermatitis.

The only apparent cause of dermatitis in the putty and file department was the epoxy resin used to repair defects in the magnesium luggage frames. Epoxy resin is not considered an allergen; therefore skin contact is probably occurring before the material is completely cured. Also, if the mix is not precise and some of the monomer is still present in the "cured" product, this monomer could be responsible for some of the sensitivity problem. The temporal characteristics of the occurrence of the rash are more suggestive of an allergic than an irritant contact dermatitis, but the high attack rate makes an allergic phenomenon less likely. Also, the distribution of the rashes, especially the absence of involvement of the hands, and the employees' work practices suggest that the rashes are more likely due to the cured resin dust generated by filing (contact) than to the uncured resin or other constituents of the resin system (allergic).

Since dermatologic abnormalities were observed infrequently at the time of the medical investigation, when weather conditions were mild, but were reported by many affected employees to be more prevalent during hot weather, heat and/or humidity apparently play an important role. In fact, some case descriptions were consistent with miliaria (prickly heat), but it does not seem likely that this could explain the majority of cases. A definitive causal relationship, based on the data available, of the dermatitis problem in this department is not readily apparent. Further evaluation by an occupational dermatologist seems warranted.

ENVIRONMENTAL

The criteria used to assess the environmental monitoring results are shown in Table 2.6. Presented are both federal regulatory requirements as well as professional guidelines as represented by the ACGIH TLV's and NIOSH Recommendations. The NIOSH Recommendations can be found in the Institute's various "Criteria for Recommended Standards" and "Current Intelligence Bulletins." In determining which exposure limit to apply, an individual strategy should be developed. For example, the first step might be to compare the measured exposure level with the OSHA PEL to determine compliance with the legal exposure limit. The exposure level can then be compared with the current ACGIH TLV, which if exceeded, may serve as a control "goal." Where the NIOSH recommendation is substantially more stringent than the current TLV, or if no TLV exists, the NIOSH value should also be used for evaluation. If no exposure limits are found, search for other professional guidelines such as might be found in the recommendations of the AIHA Committee on Workplace Environmental Exposure Levels (WEELs) or recommended by the manufacturer of the substance of concern. For each of the areas surveyed, the following conclusions were made.

HARDWARE CASTING

Air samples were collected in the casting operation for aluminum, magnesium, total particulates, fluorides, and total chlorides. The total particulates consisted of aluminum fume, magnesium fume, potassium chloride, and magnesium chloride. The maximum breathing zone sample showed a total particulate concentration of 7.0 mg/m^3, which, although less than the evaluation criteria, is a level high enough to warrant potential evaluation. The fluorides' level of 2.0 mg/m^3 was also less than the evaluation criteria of 2.5 mg/m^3, but is sufficiently close to the recommended limit to require additional

TABLE 2.6
Environmental Evaluation Criteria

Substance	Recommended Environmental Limit[a]	Source	OSHA Standard	Primary Health Effects
Aluminum fume	5 mg/m^3	ACGIH	15 mg/m^3	Necrosis of cornea
Cadmium oxide fume	0.04 mg/m^3 0.05 mg/m^3 ceiling for any 15-min period	NIOSH ACGIH	0.1 mg/m^3 0.3 mg/m^3 ceiling	Respiratory tract irritation, cough, chest pain, chills, shortness of breath, pulmonary edema, emphysema, kidney damage, anemia
Chlorides (potassium and magnesium chlorides) see nuisance particles				See nuisance particles
Chromic acid	0.025 mg/m^3 0.050 mg/m^3 ceiling for any 15-min period	NIOSH	0.1 mg/m^3 ceiling	Skin, eye, and respiratory tract irritation; skin and pulmonary allergic sensitization
Copper fume	0.2 mg/m^3	ACGIH	0.1 mg/m^3	Metal fume fever (see zinc oxide); skin and eye irritation
Fluoride	2.5 mg/m^3	NIOSH	2.5 mg/m^3	Bone changes (osteosclerosis)
Iron oxide fume	5 mg/m^3	ACGIH	10 mg/m^3	Iron deposits in lungs (siderosis) not known to be harmful

TABLE 2.6 (Continued)

Substance	Recommended Environmental Limit[a]	Source	OSHA Standard	Primary Health Effects
Magnesium oxide fume	10 mg/m^3	ACGIH	15 mg/m^3	Metal fume fever (see zinc oxide)
Nickel	0.015 mg/m^3	NIOSH	1 mg/m^3	Allergic dermatitis; cancer of lung and nasal passages
Nitrogen dioxide	1 ppm ceiling conc. for any 15-min period	NIOSH	5 ppm	Eye and respiratory tract irritation; pulmonary edema
Nuisance particles (total)	10 mg/m^3	ACGIH	15 mg/m^3	Chronic bronchitis
Phosphoric acid	1 mg/m^3	OSHA	1 mg/m^3	Irritation of skin, eyes, and respiratory tract; pulmonary edema, bronchitis
Powdered paint (mixture of titanium dioxide, calcium carbonate, polyester resin; classified as nuisance dust)	10 mg/m^3	ACGIH	15 mg/m^3	See nuisance particles

Toluene	100 ppm	NIOSH/ ACGIH	200 ppm	Skin, eye, and respiratory tract irritation; dermatitis; headache, dizziness, fatigue, weakness, drowsiness, incoordination
1,1,1-Trichloroethane	350 ppm ceiling for any 15-min period	NIOSH/ ACGIH	350 ppm	Eye irritation, dermatitis, dizziness, incoordination, drowsiness
Welding fume	5 mg/m^3	ACGIH	Standards are for individual components of the fume	Effect depends on the composition of fume (see zinc and cadmium oxide)
Zinc oxide	5 mg/m^3 10 mg/m^3 ceiling conc. for any 15-min period	NIOSH/ ACGIH	5 mg/m^3	Metal fume fever (cough, shortness of breath, weakness, fatigue, muscle and joint pain, fever, chills, sweats)

[a]Eight-hr, time weighted average unless otherwise noted.

study at a future date. Where these environmental levels are approaching the recommended values, it is appropriate to carefully consider the work load and other conditions at the time of sampling and see how these compare with average or high-volume days. The air sample for total chlorides indicated a level 0.5 mg/m^3. Calculated as potassium chloride, this would be equivalent to 1.1 mg/m^3 and calculated as magnesium chloride 0.8 mg/m^3. Both of these concentrations are less than the criterion of 10 mg/m^3.

POWERED PAINT OPERATION

The maximum air sample for total particulates collected in this area was 13 mg/m^3, as an 8-hr time-weighted average. The powered paint used is a mixture of titanium dioxide, calcium carbonate, and polyester resin. Evaluation criteria applied was that of nuisance dust, 10 mg/m^3. It is evident that the measured exposure exceeded the exposure criteria recommended by ACGIH.

BRIGHT-DIP AREA

Air samples for nitrogen dioxide, both short-term (15-min) and 8-hr TWAs, chromic acid, and phosphoric acid were collected in this area. The short-term exposure level of 2 ppm nitrogen dioxide was twice the 1 ppm level recommended by NIOSH. The 8-hr TWA value of 4.1 ppm was less than the OSHA standard of 5 ppm, but it is a level that needs further evaluation. The chromic acid level of less than 1.0 μg/m^3 was well below the NIOSH recommended environmental limit. The phosphoric acid exposure level, at one-tenth that recommended by both NIOSH and OSHA, was well below existing requirements. The main concern in this area involves the peak exposure to nitrogen dioxide, which exceeds the NIOSH recommended ceiling value of 1.0 ppm.

PLATING

The primary focus for air monitoring in the plating department involves the worker operating the barrel plater and his exposure to airborne nickel and chromic acid. The maximum exposure level to nickel was 20 $\mu g/m^3$, which, although less than the OSHA level, does exceed the NIOSH recommended environmental limit of 15 $\mu g/m^3$. In assessing the difference between the NIOSH and OSHA values, it is important to take into consideration that the recommended value is based on concern over a significant chronic health effect (cancer). The information on this health effect was developed subsequent to the adoption of the federal standard. The exposure level of 70 $\mu g/m^3$ for chromic acid (as chromium trioxide) is also less than the OSHA standard but exceeds the NIOSH recommended limit of 25 $\mu g/m^3$. It is also important to note that the sampling took place during routine operations; the greatest worker exposure is usually found when additional solution is added to the bath or when a fresh bath is made. Additional monitoring should be planned to measure these activities.

WELDING AND HARDWARE MAINTENANCE

Environmental air samples were collected under the welding hood in the breathing zone of the welder. During the survey, the welder was working with a variety of metals and welding rods; consequently, the air-sampling results shown in Table 2.5 are considered to be representative of his average exposure to toxic airborne contaminants. Of the five metals examined, the welder's exposure to two, cadmium fume and zinc oxide, were found to be excessive. The cadmium fume level was greater than that recommended by NIOSH, while the exposure to zinc oxide exceeded the OSHA standard. Exposures to copper, iron, and nickel were within environmental standards and/or recommended exposure levels.

FINAL REPAIR

The worker involved in the cleanup of the suitcases as they came through the final repair area was exposed to toluene and 1,1,1-trichloroethane. The exposure level to toluene was determined to be well below the recommended value of 100 ppm; however, the exposure to 1,1,1-trichlorethane was observed to exceed the NIOSH short-term level of 350 ppm for a 15-min period. The 8-hr TWA of 110 ppm for the solvent was well below the OSHA standard of 350 ppm.

Controls

HARDWARE CASTING

As noted in the discussion on the exposure information in this work area, it is appropriate to reevaluate the exposure levels to several of the substances found. Although all measurements for air contaminants in this area are below recommended standards, several of the measurements are at levels approaching their respective values. It is therefore suggested that additional monitoring be done in the near future, with emphasis given to both normal operating conditions as well as "worst-case" operating conditions. During the survey it was also noted that several of the exhaust ducts from the casting units were disconnected and that the ventilation system on the existing melting pots did not adequately capture the fumes being emmitted. Ventilation measurements and evaluations should be conducted prior to future air monitoring. Also, it is important to assure that routine preventive maintenance on ventilation systems be conducted and that this program be in place prior to any additional air monitoring.

PUTTY AND FILE

Several hand-washing facilities should be installed in this department. Soap used should be nonabrasive. Resin contacting the skin should be washed off immediately. All employees should wash their arms and hands at breaks, before lunch, and before they leave work at the end of the shift. Rubber gloves should be worn by the workers who handle uncured resin. Workers should also wear long-sleeved shirts or blouses and disposable paper smocks. All personnel clothing should be changed and washed daily. Rather than dry sweeping of the metal dust, a vacuum cleaner should be used to prevent the spread of metal dust in the work area.

POWDERED PAINT OPERATION

The painter in this area should wear a supplied air hood as an immediate or short-term control. This would provide both eye and respiratory protection, and by the use of supplied air hood it is more likely that the desired control will be maintained over a long period of time, unlike when a negative-pressure respirator is used. A long-term solution to this problem should involve the installation of an adequate ventilation system, *possibly* one designed after those described in the *ACGIH Industrial Ventilation Manual* for paint booths.

BRIGHT-DIP AREA

Although the NO_2 short-term exposure level is twice that recommended by NIOSH, before any major changes are made in this work area it would seem appropriate to further investigate under what circumstances these exposures are taking place. Also of importance is how often they might take place during the day, especially since the 8-hr time-weighted average value is below the OSHA standard. Consideration should be given to the position of the operator during likely peak exposures and whether or not the operator's location can be modified to reduce exposures. The use of detector tubes to obtain immediate short-term results would seem appropriate in possibly modifying the work practices of the operator in this area. If these efforts are not successful in controlling the exposures to NO_2, it would seem appropriate to consider a small, local exhaust ventilation system to reduce peak exposures. It is possible that by observing work practices we can determine that the exhaust system might only need to be turned on at those times when peak exposures are likely to occur.

PLATING

Due to the potential severity of the health effects associated with exposure to airborne nickel, it is appropriate to institute both short- and long-term controls in this area. Short-term control would involve the use of a respirator if modification of work practices does not appear feasible. For nickel dust or mist, as might be found in a plating operation, a NIOSH approved dust and mist respirator should be adequate for this exposure situation. To control exposures from the chrome plating tank, consideration should be given first to the possibility of reducing misting. Control measures might include proprietary bath additives or a layer of plastic chips on the surface of the plating solution. If control at the source of exposure does not completely solve the problem, it is often possible to increase the distance between the worker and the source through automated parts handling with remote controls. Other long-term controls might involve local exhaust ventilation, possibly designed after that suggested by ACGIH for open-surface tanks. Consideration must also be given to the various regulatory requirements contained in the OSHA standard on ventilation requirements for open-surface tanks (CFR 1910.94(d)).

HARDWARE SUBASSEMBLY

Although the causes of the rashes among subassembly employees was not determined, residual chromates on the plated parts may have been involved in some of the cases. It is therefore recommended that the parts be thoroughly rinsed before being handled by workers in the subassembly department. The wash step should be instituted in either the barrel plating area or when the parts first arrive in the hardware subassembly area.

BUFFING

Protective arm covers and gloves should be worn by the buffers.

WELDING

Since the welder may be required to work throughout the plant, as well as in his own shop area, it would be appropriate to consider the need for several control strategies to reduce his exposure to airborne contaminants. While he is in the plant, it may be appropriate for him to wear a respirator with a battery-operated air supply/filtering supplied air system. A possibility alternative approach would involve the use of a portable ventilation system that would allow the welder to provide for contaminant removal from the air whenever he works in the plant. This would reduce the exposure of other workers to welding fumes. In the welding shop it would seem appropriate to consider installation of a fixed ventilation system, such as that recommended by ACGIH for welding benches.

FINAL REPAIR

During the survey it was noted that this assembly line had a black conveyor belt that left black marks on the suitcases. The black marks, of course, were cleaned off with one of the two solvents. Other conveyor lines used colored belts that did not leave marks on the cases. It was recommended that the black conveyor belt be replaced with a colored belt, which should reduce the marking of the suitcases and consequently reduce the need for extensive use of cleaning solvents. Until the use of the solvents is decreased or engineering controls such as local exhaust ventilation can be installed, the repairman should wear a NIOSH-approved respirator for use with organic solvents.

As a final recommendation, for areas where respirators are used a respirator program involving maintenance, fitting, cleaning, and the other requirements of the OSHA standard (29CFR 1910.134) should be implemented. Plant management should arrange for the services of a dermatologist, preferably someone with experience, or at least an interest, in occupational dermatology. The continuing dermatologic problems among the workers in this plant can be very significant in reducing morale and efficiency. Therefore, the investment in a services of a dermatologist would seem appropriate in assuring that good working conditions are maintained.

REFERENCES

American Conference of Governmental Industrial Hygienists. *Industrial Ventilation*, 18th ed., ACGIH, Cincinnati, OH, 1984.

American Conference of Governmental Industrial Hygienists. *Threshold Limit Values for Chemical Substances and Physical Agents in the Work Environment and Biological Exposure Indices with Intended Changes for 1985–1986*, ACGIH, Cincinnati, OH, 1985.

Burgess, W.A. *Recognition of Health Hazards in Industry*. Wiley, New York, 1981.

National Institute for Occupational Safety and Health. *Manual of Analytical Methods*, 2nd ed., Vol. 7, DHEW (NIOSH) Publication No. 82-100, Cincinnati, OH, 1981.

National Institute for Occupational Safety and Health. *Control Technology Assessment: Metal Plating and Cleaning Operations*, National Institute for Occupational Safety and Health. DHHS (NIOSH) Publication No. 85-102, Cincinnati, OH, 1985.

National Institute for Occupational Safety and Health. *Manual of Analytical Methods*, 3rd ed., Vols. 1 and 2, DHHS (NIOSH) Publication No. 84-100, Cincinnati, OH, 1984.

Pittelkow, R. "Occupational dermatoses," in C. Zenz, Ed., *Occupational Medicine*. Year Book Medical, Chicago, 1975, pp. 191–230.

3

VAPOR DEGREASING OPERATIONS

ARTHUR B. HOLCOMB

JIMMY L. PERKINS

3.1. PROBLEM RECOGNITION

Process Description

In virtually all industrial operations, especially equipment maintenance, some form of metals cleaning and/or refinishing is an integral part. Processes vary widely from simple cleaning to refinishing using electroplating techniques. In general, the larger the operation, the more in depth the cleaning and refinishing processes. Often maintenance workers in a small facility may clean parts using a rag and a pail of solvent, while larger facilities will more likely possess a state-of-the-art vapor degreaser for the same job.

Degreasers are designed to operate within certain parameters, and specific work practices must be followed for safe and efficient operation. When these parameters and practices are not followed, degreasing often becomes inefficient, the equipment malfunctions, and/or employees become overexposed to the degreasing solvent. When degreasing units operate within specifications, resulting solvent exposures are low. These conditions must be maintained, and all other factors (e.g., air turbulence and operator work practices) should remain constant to maintain low exposure. When these criteria are not met or when exposures become excessive, the industrial hygienist must assess conditions that might have changed to affect worker exposures.

This case study describes a situation of changing conditions at a large equipment maintenance plant. Large mechanical equipment is disassembled and cleaned prior to painting and reassembly. The majority of refinished parts are from vehicular engines, transmissions, and hydraulic systems. The parts are received and degreased in wire-mesh baskets (Figure 3.1) and separated according to the type of cleaning required. In general, the parts are cleaned in the degreaser and, if necessary, paint and oxidation are removed by abrasive blasting. After cleaning, the parts are separated based on the type of protective coating required (e.g., cadmium plating or painting).

The degreaser used in this 200,000-ft^3 facility is a steam-heated, vapor-spray degreaser (Figure 3.2). The degreaser is 12 ft long, 3.5 ft wide, and 14 ft deep and is located in a deep pit, which is ventilated as shown in Figure 3.3. The degreaser is equipped with two 40-in.-long spray wands that are located at both ends of the tank and activated by the same switch. Figure 3.2 shows one of the wands outside of the tank for demonstration purposes only. Normally both wands are kept inside of the tank. The solvent used is trichloroethylene, and one 55-gal drum is consumed every wk. The degreaser holds approximately 200 gal of solvent, which fills the degreaser to a level 6

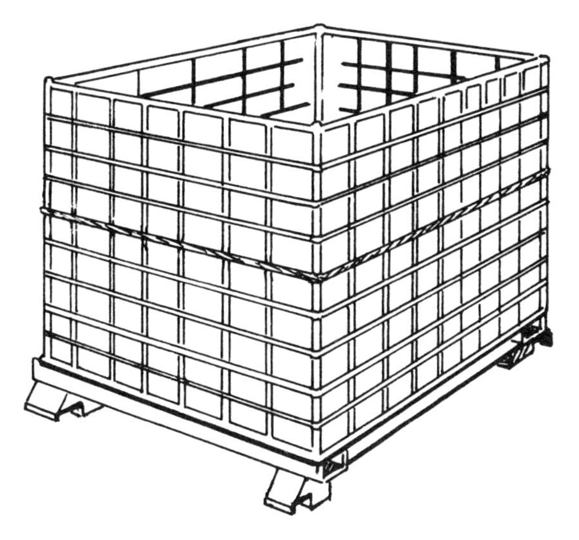

FIGURE 3.1. Drawing of wire-mesh parts basket used with degreaser.

in. above the heating coils. The ventilation hoods are at the bottom of the pit.

Condensing coils for condensation of degreaser vapor are composed of cool-water coils and refrigerated cooling coils. The condensing coils begin approximately 3 ft below the top of the degreaser tank. The catwalk is 42 in. below the top of the degreaser. The tank operating parameters shown in Table 3.1 were observed in the preliminary walk-through survey.

The degreaser operators are trained using a qualification procedure that includes a 2-day course covering proper operating techniques, use of safety equipment, and hazards associated with degreaser operations, including effects of overexposure to trichloroethylene. After completion of the course, a prospective operator is required to spend 2 wk as an apprentice to a qualified operator. Upon completion of the apprenticeship, the employee must pass a written practical examination before assignment as a vapor degreaser operator. A 1 hr,

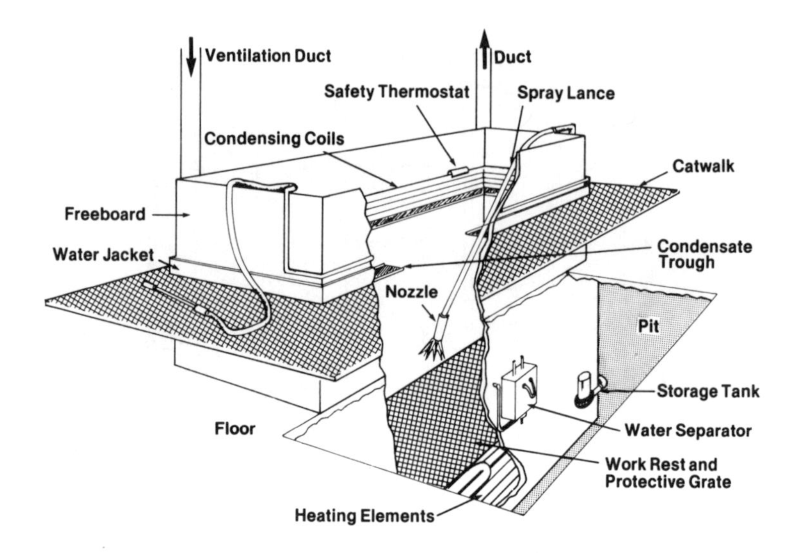

Notes: 1) Total Degreaser Height=14 feet
2) Degreaser sits in pit 8 feet deep with same dimensions as catwalk
3) Ventilation systems open at bottom of pit.
4) Catwalk is 3 feet above surrounding floor
5) Catwalk is 3 feet wide surrounding degreaser

FIGURE 3.2. Isometric view of the degreaser with cutaway view of degreaser pit and inside of degreaser.

semiannual refresher course is included in the occupational safety and health program.

Each operator is given a manual that includes the following prescribed operating procedures.

A. Start-up procedures:
   1. Turn on ventilation.
   2. Remove covers.
   3. Check solvent level.
   4. Open main water valve completely and visually check for leaks in condenser piping.

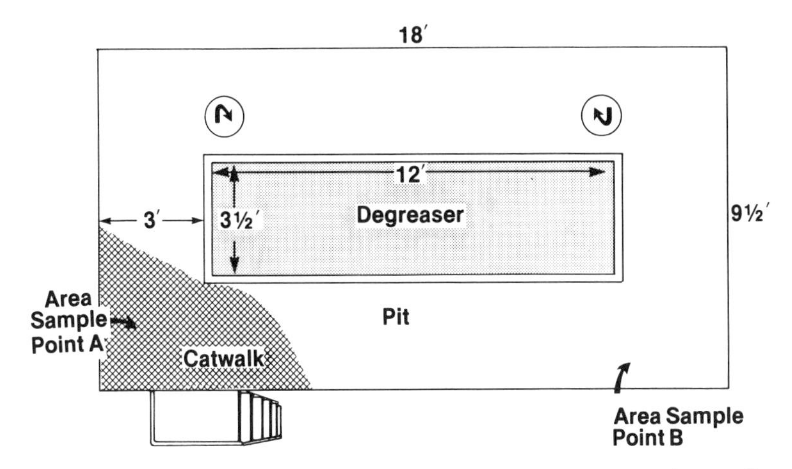

FIGURE 3.3. Top view of degreaser, pit, ventilation ducts, and catwalk.

5. Start electrical cooling unit.
6. Ensure that the return line between storage tank and boiling chamber is open.
7. Open steam trap bypass and safety valve bypass.
8. Open stem valve and flush water from lines.
9. Close safety valve bypass and steam trap bypass.
10. Check steam pressure to assure that it is between 5 and 15 psi.
11. Check thermostat adjustments.

B. Operating procedures:

1. Vapor degreasers are to be operated at a oil-solvent boiling point of 188–195°F, a steam pressure of 5–15 psi, and a vapor thermostat adjustment of 165°F.
2. Solvent level should be at least 4 in. above the heating coils and 1 in. below the work-rest platform.
3. When fresh solvent is added, it will be pumped into the degreaser, and the output end of the hose shall be

below the condenser coils during transfer of the solvent.

4. The condensation line, which can be seen on the inside of the tank, will be maintained at a level 30 in. below the top edge of the tank.
5. Excessive work loads that cause a 6-in. drop in the condensation line level, that extend outside of the degreaser, or that occupy more than 50% of the degreaser surface area will not be allowed.
6. Prevent air drafts from disturbing the vapor zone.

C. Shutdown procedures:

1. Shut off steam.
2. Allow degreaser to cool for 30 min.
3. Replace degreaser covers.
4. Close condenser water valve.
5. Shut off refrigeration.
6. Clean up work area.
7. Turn off ventilation.

D. Work practices:

1. Operate the degreaser within prescribed parameters.
2. Wear prescribed protective clothing (vinyl aprons, polyethylene gloves, and acid goggles) when spraying.
3. When starting the degreaser, always turn on condenser water before heat is applied. When shutting down degreaser, turn off heat 30 min prior to turning off the condenser water.
4. Regulate the flow of water through the condenser. The temperature of the water at the condenser outlets should be between 90 and 120°F.
5. Arrange work load on hooks, or in baskets, in such a manner as to ensure maximum exposure to the vapor and proper draining.

6. Operate degreasers so as to create minimum disturbance of the vapor.
7. Work will be introduced and removed at a maximum vertical speed of 11 ft/min.
8. When spraying is necessary, observe the following procedures:
   a. Do not start spray pump until the wand nozzle is below the vapor line.
   b. All spraying will be done below the condensation line.
   c. Angle the spray directly onto the work and in a manner as to prevent splashing of the liquid solvent.
   d. After spraying, ensure that the nozzle drains completely before it is removed from the tank.
   e. Remember that spraying cools parts and causes additional condensation.
9. Degrease only nonporous material that is cool and dry.
10. Regardless of the combination of treatments that may be used, the final phase of the degreasing process shall be the suspension of the work in the vapor until the part reaches the vapor temperature. At this point, condensation of solvent on the part will stop. This provides for maximum cleaning and assures proper drying. The thicker and heavier the material being cleaned, the greater the amount of time required for this step.
11. After the parts are cleaned and condensation has stopped, raise the basket above the vapor line and allow the material to dry and cool and drain in areas where pooling has occurred prior to handling. The material is hot when removed from the vapor and should not be immediately touched.

E. Special safety precautions:
   1. Repeated skin contact with liquid or concentrated vapor can produce dermatitis.
   2. Overexposure to vapor can cause drowsiness, giddiness, dizziness, vertigo, fatigue, headache, nausea, and vomiting. If any of these symptoms appear, immediately notify the supervisor.
   3. If the body is splashed, immediately remove soaked clothing and wash exposed area thoroughly with running water. Do not put solvent-wet clothes on again.
   4. If eyes are splashed, first flush eyes thoroughly with large amounts of water and then obtain medical attention.
   5. If solvent is swallowed, induce vomiting—unless the person is unconscious—and call a physician.
   6. If a person faints or loses consciousness due to inhalation of vapor, remove the victim to fresh air and obtain medical attention at once. If breathing stops, start artificial respiration at once.
   7. Solvent vapors are hot and will cause burns.

Due to the time required for start-up and shut-down work procedures, operation of the degreaser is performed for an average of 6–7 hr per day. Approximately 15 loads of work are cleaned during that period. A degreasing routine consists of the basket of parts being lowered into the vapor zone, waiting for the parts to reach the temperature of the vapor, spraying the parts with cool solvent, waiting for the parts to reach vapor temperature again, and removal of the parts from the tank.

The operator is in the immediate vicinity of the degreaser tank when a load enters the degreaser, when spraying the parts, and when removing parts from the degreaser, but not during the waiting period. There are four persons who are qualified to operate the degreaser, but usually no more than two are performing degreaser work at any one time. The others will be at different tasks, on leave, or at lunch. When not required at the tank, at least one operator is busy in the preparation area readying parts for cleaning or for movement to other areas.

As part of the plant medical surveillance program, operators are given a preplacement and an annual physical that includes liver function test, urinanalysis, determination of cardiac status, and pulmonary function test.

Nature of the Problem

In this case study, you are a corporate industrial hygienist for a facility that refurbishes large construction equipment and are called upon to investigate a grievance filed by several degreaser operators. Following union grievance procedures, they have requested differential pay for exposure to hazardous substances. You are expected to evaluate the hazards involved in the job and recommend any actions necessary to correct unhealthful working conditions. As an industrial hygienist, it is your responsibility to determine if proper operating conditions are being met and if exposures are excessive.

Questions

1. What specific work practices will you be interested in observing?
2. What operational parameters should be of interest? Give particular attention to the design of the degreaser and how it *should* perform when used properly. What is proper?
3. How will you evaluate the contribution of different operations to the overall trichloroethylene exposure?

4. You *should* assume that process changes will be necessary and that a study of conditions before and after the changes are implemented will be necessary. How will you conduct the studies?

3.2. EVALUATION RESULTS

Preliminary Findings

A comprehensive evaluation of a vapor degreaser should include evaluation of equipment-operating parameters, operator's work practices, and determination of exposure to toxic chemicals. The equipment-operating parameters are in many cases subject to change and should therefore be checked both at the start of operation and at the conclusion of each day's work. To facilitate evaluation, the operating parameters for this study will be considered constant and have been previously listed in Table 3.1.

In observing work practices, the industrial hygienist should look for deviations from prescribed practice and other situations that may contribute to worker exposure to hazardous conditions. In this case, observations should concentrate on operator procedures during the time when the operator is in

TABLE 3.1
Equipment Operating Parameters

Parameter	ASTM Recommended Standard	Actual
Hoist speed	<11 ft/min	18 ft/min
Solvent-oil boiling point	188–195°F	190°F
Steam pressure	5–15 psi	10 psi
Water jacket temperature	100–110°F	110°F
Condensate temperature	<164°F	160°F

the immediate vicinity of the degreaser tank. The following observations were made.

1. To compensate for the excessive hoist speed, the operator lowers parts into the degreaser tank by starting and stopping the hoist. As parts enter the vapor zone, vapor turbulence is created.
2. When large parts (e.g., engine cylinder heads) or baskets containing a large number of small parts are lowered into the degreaser, the condensation line drops approximately 2 ft. To conform to prescribed procedure, the operator lowers these parts into the vapor until the top of the vapor zone (condensation line) drops 6 in. and stops the hoist to allow the vapor zone to return to the normal operating level. This procedure is repeated until the parts are completely submerged in the vapor zone. Throughout this procedure the operator is in the immediate vicinity of the degreaser tank. The repeated starting and stopping of the baskets results in turbulence in the vapor zone.
3. The vapor degreaser is equipped with two spray wands operated by a single pump. This results in both wands spraying when the pump is turned on. One wand is used by the operator, while the other is suspended on the side cf the degreaser tank with the discharge directed into the tank.
4. Due to the freeboard distance of 3 ft and the spray-wand length of 40 in., the parts must be raised to the top of the vapor zone in order for parts to be effectively sprayed.
5. The switch for the spray wands is located at one end of the degreaser. The operator is normally located in the center of the degreaser. On many occasions the operator neglected to turn the pump off. This procedure re-

sulted in both wands continuously spraying and creating a constant turbulence in the vapor zone.

6. The rails on the bottom of the parts baskets are concave upward. These troughlike rails are positioned in a manner that allows for the accumulation of solvent.
7. When removing parts from the degreaser, the operator frequently did not allow the parts to stay within the freeboard area until the parts were dry. In addition, some parts accumulated trichloroethylene due to their shape. The liquid did not evaporate during the drying process and usually ended up on the floor during the unloading process.
8. When the degreaser pit ventilation system is operating, an updraft is created on the end of the degreaser, which is adjacent to the push side of the ventilation system. It appears possible that if any trichloroethylene vapor is located in the bottom of the degreaser pit (possibly due to leaks or spills) it will be blown into the operator's area.
9. A review of the degreaser operators' medical histories with the plant physician revealed that one operator had a forced vital capacity of less than 50% and that another had a history of cardiac difficulties. The first worker had removed the exhalation valve from his respirator in order to breathe more freely.
10. Operators are provided with polyethylene gloves, vinyl aprons, chemical splash goggles, and chemical cartridge respirators. Use of gloves, aprons, and goggles is mandatory. Use of respirators is optional. One operator stated that he used his respirator when the odor of trichloroethylene was "overpowering." During the survey, respirators were frequently worn.
11. Solvent vapors could be smelled at the platform or catwalk and in the staging or parts preparation area.

In-Depth Studies

In the final analysis, the best evaluation of any set of controls or procedures for a hazardous situation is the amount of exposure of personnel to the toxic agent involved in the operation. In this case, exposure to trichloroethylene can be determined by several different methods. There are several direct-reading instruments available for measuring trichloroethylene. However, due to the nature of the exposure profile (frequent and rapid changes in concentration), the wide range of concentrations occurring, and the fact that we are primarily interested in a measure of the workers' 8-hr, time-weighted average (TWA) exposure, use of direct-reading instruments would be difficult, even with a computerized integrator. However, these instruments can be invaluable in determining those operations that produce high peak exposures and in evaluating the result of changes in work practices or equipment-operating parameters. In this case, a MIRAN Model 103 Infrared Analyzer was used to evaluate exposure profiles as operating parameters were changed.

Personal monitoring is the most useful method of determining (TWA) exposures. In recent years, use of passive dosimeters for personal monitoring has gained support within the industrial hygiene community. Compared to the convential charcoal tube and vacuum pump method, dosimeters are more convenient to use and have been proven to provide comparable results under many conditions. At least one study has demonstrated the efficiency of passive monitors in evaluating exposures to degreasing solvents.

In this case study, samples used to calculate the TWA exposure were collected using organic vapor passive dosimeters. Short-term exposures (less than 15 min) and area concentrations were determined using a charcoal tube and a constant flow pump, calibrated at a flow rate of 200 mL/min. For the 8-hr TWA concentrations, consecutive 2-hr personal samples covering the day's operation were collected for 5 consecutive days on each employee. For each measurement reported in Table 3.2, two employees were sampled, and the highest concentration for the pair is reported. Admittedly this will overestimate the usual 8-hr TWA for each employee but should be more representative of a worst-case 8-hr TWA than the average of the two measurements.

The three major sources of exposure (loading, spraying, and unloading), were sampled for their duration, which was usually less than 15 min. Only a portion of the approximately 15 loads per day were monitored for these three processes. Results of the analysis of samples are provided in Tables 3.2 and 3.3.

In evaluating the results of short-term exposures, it is essential that notes be taken to allow the industrial hygienist to relate exposures to a particular operation. In this study, a short-term sample was started when the operator entered the vicinity of the degreaser tank and ended when the operator exited the area.

In addition to integrated personal and area samples, a MIRAN Model 103 was used to evaluate the effectiveness of process and work practice changes on a real-time basis. Al-

TABLE 3.2
Worst-Case Results of 2-Hr Consecutive Samples (n = 2) Using Passive Dosimeters (PPM, Trichloroethylene)

	Monday	Tuesday	Wednesday	Thursday	Friday
	84	74	100	60	49[d,e]
	90	120	58[a,b]	38[c]	36
	124	92	70	48	44
	96	78	72	45	48
8-hr TWA	99	91	75	48	44

[a–e]Process and work practice changes were made. These changes, which will be described in Section 3.3, were permanent starting with the time period indicated.

TABLE 3.3
Results of Short-Term Samples of Approximately 25% of All Degreaser Loading, Spraying, and Unloading Operations Using Charcoal Tubes (PPM, Trichloroethylene)

Loading		Spraying		Unloading	
Min	Concentration	Min	Concentration	Min	Concentration
		Monday			
4	122	5	190	5	161
5	135	7	224	4	170
4	140	6	201	7	150
6	108	4	220	4	163
5	150	5	114	7	122
		6	214	6	141
		5	218	5	140
		10	160[f]		
		Tuesday			
4	124	5	201	5	160
4	132	3	198	3	172
7	130	7	224	7	150
5	110	9	209	7	148
6	113	13	162[f]	7	163
		4	201	4	171
		6	198		
		Wednesday			
4	130	8	190	5	198
7	112	7	202	5	122[b]
2[a]	150	10	204[f]	7	102
2	160	5	198	6	99
1	175	6	160	6	108
2	130	6	198	4	110
		7	201	7	101
		5	182		

TABLE 3.3 (Continued)

Loading		Spraying		Unloading	
Min	Concentration	Min	Concentration	Min	Concentration
		Thursday			
2	170	5	198	7	123
2	172	4	155[c]	5	102
2	138	5	142	5	112
2	155	6	129	6	113
1	134	6	148	8	103
2	170	7	121	6	118
		9	131	5	195
		5	129		
		Friday			
2	173	5	126[d]	8	98[e]
2	173	7	102	6	107
2	170	6	121	4	97
2	135	4	98	6	87
2	132	5	118	4	98
2	170	7	95	4	110
		6	108	8	101
				8	95

[a-e]Indicate that process and work practice changes were made. These changes, which will be discussed in Section 3.3, were permanent starting with the time period indicated.
[f]Both spraying and unloading were performed.

though personal samples can certainly be used to show the effectiveness of these changes, results are usually not available for some time. On the other hand, real-time analysis provides an immediate measure of effectiveness. The real-time data are not reported; however, changes were not instituted nor followed up by personal sampling unless an examination of the real-time data indicated a probable reduction in exposure. Personal sample results, as indicated in Tables 3.2 and 3.3, reflected the high effectiveness of process changes in lowering exposure levels.

Finally, area samples for trichloroethylene were taken at locations A and B, as shown in Figure 3.1. The results indicated that the airborne levels of trichloroethylene at A were four times those at B.

Questions:

1. Based on these findings, what process and work practice changes do you recommend?
2. Do you have any long-term recommendations for monitoring the performance of the degreaser?
3. What, if anything, would you recommend with respect to medical examinations and the use of protective clothing?

3.3. CONTROL STRATEGIES

Short-Term Controls

The following process and work practice changes were made during the course of this study. The letter identifying each change relates to the footnotes in Tables 3.2 and 3.3.

(a) As noted earlier, operators had to lower the basket into the degreaser in stages in order to allow the vapor level to remain stable. This caused workers to remain in the pit area for long periods of time. This procedure was corrected by instructing the workers to completely lower the basket in one step and then immediately leave the area. As can be seen in Table 3.3, exposure levels of trichloroethylene during loading may have increased somewhat because of this change, but exposure time decreased considerably. At the same time, the hoist speed was corrected to the ASTM Standard of 11 fpm, and the higher alternative hoist speed was disconnected. It was thought that this step would reduce turbulence and therefore exposure levels but would not cause a considerable increase in exposure time. Consequently, exposure time for this process decreased from an average of 5 to an average of 2 min. Approximately 1 min was needed for the parts basket to descend; the other minute was needed for loading the parts basket onto the hoist and in preparation of the parts basket.

(b) Workers were instructed to raise the basket into the freeboard area after parts were dry and clean. Using the hoist, they moved the basket to the end of the degreaser and propped one end on the ledge above the condensing coils. The hoist was then lowered slightly to allow the basket to tilt. This practice was used for parts that tended to pool liquid solvent, and it allowed those parts to drain and further dry. After propping the basket, the operator left the area for 5 min to allow further drying. While this may have increased unloading time, it seems to have reduced exposure levels during unloading by about one-quarter (see Table 3.3).

(c) The degreaser had two spray wands that worked in unison. Flow of solvents from the wand not in use caused turbulence of the vapor in the degreaser and led to increased loss of solvent vapor. Since this wand was never used, it was removed from service. This action seems to have reduced exposures during spraying.

(d) The remaining spray wand was extended by 2 ft. This allowed the operator to spray parts at a location deeper in the degreaser without having to raise and lower the basket.

(e) Since the rails on the bottoms of the baskets (see Figure 3.1) were concave up, they collected solvent. This was remedied by drilling several holes in the rails at 2-in. intervals to allow for drainage.

One final change was made which should have affected the 8-hr TWA values. As noted earlier (see Figure 3.3), a push - pull ventilation system was in use. Area samples indicated that the push side (A) apparently tended to cause higher vapor concentrations than on the pull side. After the push fan was turned around and made to pull, concentrations on both sides of the degreaser pit were reduced and were approximately equal.

The overall result of these changes is best shown in Table 3.2. Eight-hour TWA concentrations of trichloroethylene were reduced from almost 100 ppm to 44 ppm. The current OSHA permissible exposure limit for trichloroethylene is 100 ppm, with a 200 ppm, 5 min ceiling in any 2-hr period and 300 ppm peak. The current ACGIH-TLV is 50 ppm and has a 15-min 200 ppm ceiling. Consequently, while the old exposure levels were near the OSHA standard, the new exposure levels are near the ACGIH TLV.

Long-Term Controls

As described in Section 3.2, a review of medical histories revealed that at least one worker had medical problems that would preclude the use of a respirator. If respirators are to be used, a respiratory protection program as required by OSHA, 29CFR1910.134 must, at a minimum, be instituted. As part of this program, training should emphasize that exposure levels have been lowered to around 50 ppm, 8-hr TWA, and therefore respirators are not required. Any employee who wishes to use a respirator should be allowed to do so, provided that it is used correctly. Realistically, it may be difficult to convince management to allow the voluntary use of respirators. However, even though in compliance with the OSHA PEL, the IH should consider ACGIH (50 ppm) and NIOSH (25 ppm) recommendations and variations in individual sensitivities in allowing voluntary respirator use. One should always remember that ACGIH TLVs are designed to protect *nearly* all workers, as stated in the first paragraph of the introduction to that document. In any case, workers should be trained and medically capable of using respirators. Use of respirators is practically useless, and in some cases harmful, unless these minimum requirements are followed.

It should be realized that only one 8-hr sample (four 2 hr samples) was used to arrive at the final exposure level on Friday (see Table 3.2). It seems probable that the reading is accurate, based on levels of previous days, levels from short-term samples, and corrective actions. However, sampling on 1 day will not suffice as an adequate record of exposure profiles. Consequently, in the future, sampling should be conducted on other days. These may be chosen on a random-statistical or on a suspected worst-case basis.

Since trichloroethylene may adversely affect the cardiac system, it would be prudent that the worker with cardiac problems should probably not be exposed to trichloroethylene at the concentrations initially found. Consideration of additive or synergistic effects from exposures to other agents should be considered. For example, trichloroethylene has some toxicological properties similar to ethanol. In particular, this worker and all workers should be asked about drinking habits, as ethanol also effects the liver. Removal of this employee from the job requires many considerations such as a review of union contracts and nearness of employee to retirement. This decision should be made upon advice from one or more physicians.

It appears that three administrative areas are ineffective. First, workers with cardiac and/or respiratory problems should be more thoroughly screened during preplacement or periodic physicals. Liver, cardiac, pulmonary, and kidney systems should be examined. Baseline data on function of these vital organs and the physician's recommendations should be used for making decisions about placing new employees or removing present employees from a job. Criteria for these actions should be documented, so that all employees are treated as objectively and fairly as possible. The medical program should be reviewed to see that it follows these guidelines.

The second administrative weakness involves training. Based on the findings of the study, it seems likely that training is not effective and is probably not taken seriously by either management or labor. Workers should be trained in the proper operation of a piece of machinery; they should report any suspected malfunctions such as fast hoist speed or pooling of solvent in parts baskets. They should be taught to have respect for chemicals they handle. Also, they should be encouraged to make recommendations on how working conditions can be improved, as they are usually most familiar with those conditions. Furthermore, training in respiratory protection should be such that a worker would realize that removal of the exhalation valve from a respirator makes it absolutely useless.

The third area involves the apparent lack of monitoring of air levels. As stated at the outset, degreasers should cause low exposures when properly operated. However, changes in work practices, operations, personnel, and even weather can cause differing exposures. Therefore, exposures should be monitored periodically. In this case, at least yearly would be recommended.

Alternative Solutions

As is usually the case in industrial hygiene, several solutions may be possible, and any that work are correct. In the case of the worker with reduced respiratory capacity, a positive-pressure air-line respirator could have been provided to allow him to wear a respirator more easily. However, the air line itself can be hazardous or unsafe. A self-powered air-purifying respirator would be another alternative.

Another recommendation consistent with good industrial hygiene practice is to substitute a less toxic chemical. In addition to trichloroethylene, four solvents commonly used for vapor degreasing are trichlorotrifluoroethane, methylene chloride, perchloroethylene, and methyl chloroform. Of the four solvents, the solvent with the highest PEL is trichlorotrifluoroethane and for several reasons, including low toxicity, it is a good substitute (see, for example, Burgess). However, it is also the least used of the five solvents because of its cost and lower boiling point. Methyl chloroform has the next highest PEL. Trichloroethylene has the lowest PEL. Considerations such as the type of surface to be cleaned or nature of the material to be removed, temperature required for cleaning, and cost of a different solvent must be investigated prior to the substitution of a new solvent.

In many vapor degreasing situations, industrial hygienists have recommended the use of methyl chloroform. Although the PEL of methyl chloroform is currently higher than for the other commonly used solvents, methyl chloroform does have environmental properties (a long atmospheric half-life) that may cause it to be more strictly regulated by the Environmental Protection Agency. Furthermore, there is no assurance that the PEL of methyl chloroform will not be lowered. If a substitution is made, certain engineering changes probably would have to be made to the degreaser. A water separator is required for all degreaser solvents; however, for methyl chloroform the use of a water separator is critical. In the presence of excess water vapor, methyl chloroform reacts to form hydrochloric acid, causing damage to the parts being cleaned and to building structural material. Stabilizers can be used to slow this process; however, one should always be aware of the stabilizer being used, the concentration in use, and its toxicity.

For vapor degreasers, as with many other pieces of equipment, operating parameters are the major factors controlling operator exposure to the solvent. The ideal situation would be a remote control operation, with minimum exposure to the operator. These types of controls are possible with current technology; however, their cost may prohibit their use in a small plant.

Ventilation control of degreasers is widely debated among industrial hygienists. In the example presented here, the industrial hygienist was fortunate in that he tried all of the obvious solutions first and they worked. In some cases this does not occur, owing to the nature of the solvent, the size of parts cleaned, or other operating parameters. In these situations, the remaining solution may well be ventilation, and it should be used if necessary.

There are at least two administrative measures that should be initiated. The first involves the use of an operator's daily log. The log should include operating parameters, work-load conditions, and observations. Second, a material balance should be performed on the operation. The log should contain dates on which new solvent is added. When the amount of solvent added increases, it usually means that conditions have changed, and exposures are increasing.

In making recommendations to correct exposures, the industrial hygienist must consider the benefit received versus the cost. In most cases, management will make inexpensive changes, even though exposures are below permissible levels. On the other hand, very expensive or difficult-to-administrate changes will usually be opposed without the demonstration of some real benefits, such as reducing exposures below the legal limit, or reducing operating expenses, compensation costs, or sick leave.

REFERENCES

Burgess, W.A. *Recognition of Health Hazards in Industry*, Wiley, New York, 1981, pp. 20-32.

Mazur, J.F., D.S. Rinehart, G.G. Esposito, and G.E. Podolak. Evaluation of passive dosimeters for assessing vapor degreaser emissions, *Am. Ind. Hyg. Assoc. J.* 42:752-756, 1981.

Title 29, *Code of Federal Regulations* (CFR), Part 1910, Occupational Safety and Health, Government Printing Office, Washington, DC, 1983.

American Conference of Governmental Industrial Hygienists. TLV's® *Threshold Limit Values for Chemical Substances and Physical Agents in the Workroom Environment with Intended Changes for 1983 - 84*, ACGIH, Cincinnati, OH, 1983.

Wheeler, J.B. et al., eds. *Handbook of Vapor Degreasing*, American Society for Testing Materials (ASTM) Special Technical Publications No. 310A, Philadelphia, 1978.

4

MAINTENANCE OF SEPARATORS IN THE PETROLEUM INDUSTRY

JIMMY L. PERKINS

4.1. PROBLEM RECOGNITION

Process Description

In the petroleum industry, process waste water passes through several treatment stages before it may be discharged to a natural body of water. One of the early stages in this water pollution control process involves use of the principle that water is heavier than oil and most of the hydrocarbons derived from oil. To make use of this principle, oily waste water is slowly pumped through a large rectangular settling basin (see Figure 4.1). In this case study the basin is open to the atmosphere, although more separators are being covered in order to prevent fugitive emissions of volatile organics. Flow is controlled so

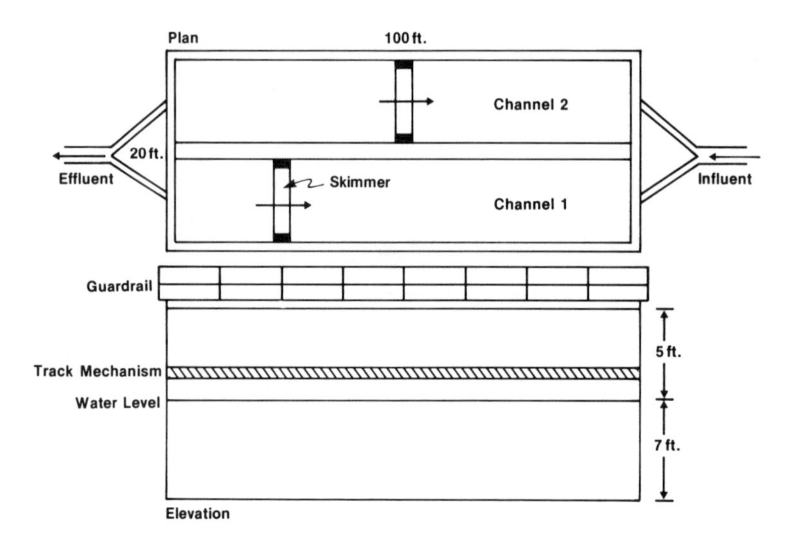

FIGURE 4.1. Schematic of a typical oil/water separator (drawn from data of H.R. Jones and American Petroleum Institute).

that a large percentage of the oil/water mixture is allowed to separate; the oil rises to the top of the water column.

A skimmer pipe, which traverses the width of the basin, moves the length of the basin and rides just below the waste water surface. The skimmer is attached to a track on either side of the basin, and is often propelled by a chain and sprocket assembly.

Oily waste separated from the waste water is pumped to a holding tank. The remaining waste water usually goes to an aeration treatment system. Typically, the temperature of the influent is 90–110°F.

Only certain types of refinery wastes enter the separator. These include waste water from the oil desalinization process, oily water drawn from the bottom of storage tanks, water condensate from steam product strippers, pump gland cooling water, and water used to wash down equipment spills. The latter can change the complexion of the influent to the separator most drastically, since spills are not continuous but, rather, random and discrete events.

On a continuous basis, the influent to the separator will usually contain a mixture of hydrocarbons not unlike the crude oil being processed. However, surges or slugs of concentrated single hydrocarbons can occur. This is particularly true if specialized units for distilling hydrocarbons such as benzene, toluene, and xylene are present at the refinery. Also, hydrogen sulfide and other sulfur compounds may enter the separator in discontinuous patterns.

Nature of the Problem

Of concern in this case study are the maintenance workers who usually enter the separator three to four times per year to repair the skimmer track and chain mechanism or for other maintenance. Typically, 4–16 hr are necessary for these tasks. The workers work from a platform that is suspended by a hoist like that used by window washers, generally 1 to several feet above the oily water.

The refinery involved in this problem is located in South Texas along the Gulf Coast. The problem involving the separator arose in July during a maintenance operation involving three workers. After a period of 2 hr (1:00–3:00 p.m.) in the separator, all three workers began to feel dizziness, fatigue, and headache. They had been squatting on the platform while performing the maintenance task, and upon standing, one worker fainted. Of course, this halted the operation. Medical help was sought from the on-plant physician, and a safety specialist performed an accident investigation.

As the industrial hygienist, it is your job (1) to ascertain if work can continue on the morning of the following day, and (2) to perform a long-term study to determine the probability of hazardous atmospheres in the separator. You arrive on the scene about 30 min after the incident and find the workers have been taken to the clinic, and a safety person is also at the clinic obtaining information.

Questions

1. Based on the information given, what factors and conditions do you believe caused the incident?
2. In order to determine if work can continue at the separator on the following day, what industrial hygiene equipment might you use? What will be your approach?
3. What reports or documents should you review which involve the accident, and what personnel, if any, should you interview?
4. What approach will you use to conduct the long-term study of the separator?

4.2. EVALUATION RESULTS

Preliminary Findings

The best potential sources of information for this study are:

Accident investigation report

Physician's report

Weather data (use on-site data, if available, or call the nearest weather station)

Work permit data (at many facilities work permits are issued by a safety representative prior to entrance into any vessel)

Worker interviews

The industrial hygienist determined from the plant weather station that the temperature from 1:00 to 3:00 p.m. had ranged from 101–104°F. Relative humidity varied from 55 to 57% over the period. The weather station reported that the temperatures for the next day were to vary from 78–105°F. Relative humidity was expected to peak during the coolest portion of the morning at about 85%.

The work permit report revealed that combustible gases had been checked with a Gas Tech model 1641 Combustible Gas Indicator, prior to work start. (This instrument can also be used to monitor O_2 and H_2S.) A concentration of less than 2.5% of the LFL (lower flammable limit) had been found. Although the meter needle did apparently move, meter divisions are only at 5% intervals. Therefore, one should only read to the nearest 2.5%.

The detector had been calibrated against methane, which has an LFL of 5% v/v. Since the Gas Tech 1641 was calibrated for methane, the LFL reading for mixed refinery vapor would not be exactly correct. Hexane, heptane, or octane are probably better for calibration, since these compounds are more representative of common derivatives of crude oil. Gas Tech states that for an analyzer calibrated with methane, but being used in a hexane atmosphere, the results should be multiplied by 2.5 to account for hexane's greater combustibility. The LFL of hexane and other refinery chemicals are shown in Table 4.1. H_2S was not detected. The O_2 level was 20.5%.

The physician's medical report revealed that all three workers had rapid pulse rates. Symptoms reported were dizziness, fatigue, headache, in all three, and slight nausea in the worker who fainted. All patients complained of eye irritation and slight irritation of the respiratory tract. All three patients recovered rapidly after being administered normal saline solution intravenously. A diagnosis of heat exhaustion was made by the physician.

TABLE 4.1
Lower Flammable Limits for Some Petroleum Constituents and Products

Compound	v/v of Air %
Methane	5
Butane	1.5–1.65[a]
Hexane	1.2
Heptane	1.1
Octane	0.8–0.95[a]
Benzene	1.2
Toluene	1.2–1.27[a]
o-Xylene	1
Ethylbenzene	1
Gasoline	1.3
Kerosene	0.6

[a]Ranges indicate values taken from two sources (Draeger and American Conference of Governmental Industrial Hygienists) and show that values are not absolute.

The accident report indicated that each worker had been wearing a half-mask respirator with organic vapor cartridges. They also wore Coast Guard-approved life vests. The workers were 21, 22, and 35 years of age. The worker who fainted was 35 and was a new employee (less than 6 mo service). The others had worked as maintenance personnel for more than 2 yr.

All workers at the plant are trained in respirator use and receive a qualitative fit test upon employment. Thereafter, annual qualitative fit tests and training sessions are performed by safety personnel.

Approximately 45 min after the accident occurred, the industrial hygienist entered the separator on the platform. He wore a full-face respirator with a supplied-air line. He carried a detector tube pump and detector tubes. Average results from two of each detector tube type were as follows:

Benzene 5/a	5 ppm (precleanse area turned completely brown)
Toluene 5/a	125 ppm
Hexane 100/a	275 ppm
Hydrocarbon 2	375 ppm

In addition, a TLV Sniffer (Bacharach Instruments) was used and results indicated 700 ppm. The instrument had been calibrated with hexane.

In-Depth Studies

During the month following the accident, the industrial hygienist took 2-hr air samples from the separator using large size (200/400 mg) charcoal tubes and low flow pumps (calibrated at 50 cm^3/min). The MSA Company reports that standard 50/100 mg charcoal tubes have a capacity of roughly 15 mg total organics or 10 L of air pumped through at 500 ppm total organics (assuming 100% absorption and an average molecular weight of about 75). Since 6 L of air were to be collected and total concentrations were unknown, but possibly greater than 700 ppm, a 200/400 mg tube was chosen to avoid breakthrough to the back-up section. A 2-hr sampling time was selected as optimum in terms of loading and for administrative reasons.

Twenty-four sample periods of 4 hr duration were selected at random from a 28-day period as follows. Each day was divided into three 8-hr shifts. Eight dates for each shift type were drawn at random, and one of three 4-hr time periods was drawn at random from each 8-hr period. The three different 4-hr time periods possible for each 8-hr shift were the first 4 hr, the middle 4, and the last 4. Two consecutive 2-hr samples were taken in each 4-hr period selected.

Analysis of samples was made by gas chromatography for

TABLE 4.2
Results of Analysis (ppm) for Eight Hydrocarbons in 24 Four-Hour Sample Periods[a]

Compound	Sample No.							
	First Shift (8:00 a.m. to 4:00 p.m.)							
	1	2	3	4	5	6	7	8
Hexane	7	7	12	16	9	9	7	8
Heptane	7	12	13	20	10	11	12	13
Octane	10	11	12	30	10	9	9	8
Benzene	ND[b]	ND	ND	3	ND	ND	1	ND
Toluene	ND	1	ND	3	ND	2	ND	1
Xylene[c]	ND	1	ND	5	1	ND	ND	2
	Second Shift (4:00 to 12:00 p.m.)							
	1	2	3	4	5	6	7	8
Hexane	8	9	20	14	14	8	7	12
Heptane	7	7	33	10	16	12	12	11
Octane	12	11	20	10	14	7	12	10
Benzene	ND	ND	2	1	3	ND	ND	ND
Toluene	ND	ND	3	2	3	ND	1	1
Xylene[c]	ND	8	8	ND	5	3	2	4
	Third Shift (12:00 p.m. to 8:00 a.m.)							
	1	2	3	4	5	6	7	8
Hexane	7	12	8	13	7	17	8	10
Heptane	1	20	10	11	12	14	10	11
Octane	8	8	9	8	12	20	10	11
Benzene	ND	ND	ND	ND	3	ND	ND	ND
Toluene	ND	1	ND	2	111	5	2	3
Xylene[c]	ND	3	4	3	97	9	ND	8

[a]Results are averages of two consecutive 2-hr samples.
[b]ND, not detected (approximately < 1 ppm).
[c]Sum of *o*, *p*, and *m* isomers.

hexane, heptane, octane, benzene, toluene, ethylbenzene, and *o*-, *p*-, and *m*-xylene. Results are shown in Table 4.2. These compounds are the most prominent alkanes and aromatics in crude petroleum, as shown in Table 4.3. Alternatively, the industrial hygienist could check records for the type of crude oil being refined and obtain from the records of the laboratory a more reliable analysis of its constituents and their concentrations.

Questions

1. Based on the findings, should worker entrance to the separator be allowed the following day? State your rea-

TABLE 4.3
Hydrocarbons Isolated from a Representative Sample of Crude Petroleum, Ponca City, OK[a]

Compound	Percent Value
n-Hexane	1.9
n-Heptane	2.3
n-Octane	1.9
n-Nonane	1.8
93 other cyclic and branched alkanes	<1.8 each
Benzene	0.15
Toluene	0.51
o-Xylene	0.27
p-Xylene	0.1
m-Xylene	0.51
36 other benzene and napthalene derivations	<0.51

[a]Adapted from Environmental Protection Agency, 1977.

soning. If entrance is allowed, what precautions must be taken?

2. Based on the long-term study, do you think that hazardous conditions can exist in the separator with respect to hydrocarbon vapors? What other studies should or could have been undertaken by the industrial hygienist? Do you have any criticism of the structure of the long-term study?
3. What recommendations would you make to management regarding future worker entrance into the separator? Assuming that all situations that led to the incident, or that may have at least indirectly caused the incident, have been presented in this chapter, would you have any recommendations regarding plant wide safety and health practices?

4.3. CONTROL STRATEGIES

Discussion

There is no other way to maintain the separator than to enter on a platform as described earlier. Maintenance is required about four times per year. Results from the various tests performed would indicate that respirators are necessary. Of course, the difficulty with this determination is that a mixture of hundreds of hydrocarbons are present, and one can only estimate the toxicity of the mixture based on environmental limits of other similar mixtures or major components of the separator mixture.

The worker health issue of greatest probable concern in this case study was the unusual heat that was present on the day of the accident. The on-site weather station indicated 105°F and 55% relative humidity at the time of the accident. Although there is no OSHA standard for heat stress, both NIOSH and the ACGIH recommend that when the work is performed outdoors with a solar load the WBGT Index be used to predict potential harmful situations:

$$\text{WBGT} = 0.7\ \text{NWB} + 0.2\ \text{GT} + 0.1\ \text{DB}$$

where NWB is the normal wet bulb temperature, GT the globe temperature, and DB the dry bulb temperature. The WBGT Index and an estimate of the work load are combined to determine an appropriate work–rest regimen.

Of the above variables, only the dry bulb is known in this case study (105°F). NWB is the temperature produced when water evaporates from a wick placed over the bulb of the thermometer. As the wind speed increases, NWB decreases. The minimum NWB temperature on this day would be that occurring if the wind speed were maximized or if the thermometer were slung rapidly through the air, as with a sling psychrometer. From a psychrometric table, one can deduce that at 105°F and 55% relative humidity the *minimum* wet bulb temperature would be 90°F.

In further estimation, if one also minimized the effects of solar radiative heat and assumed that GT = DB, then WBGT = 0.7 (90°F) + 0.3 (105°F) = 94.5°F or about 35°C. According to ACGIH, only well-acclimatized workers, under medical surveillance, should be exposed to this level of heat stress. Furthermore, the work load should be light and rest breaks should be provided during at least 75% of each hour. Actually, the situation is even worse than calculated, since the globe temperature was undoubtedly higher than dry bulb, and the wind velocity probably did not exceed 5 mph. Therefore, the NWB was certainly higher than 90°F.

Consequently, it is not surprising that the worker who fainted had the least on-time duty and presumably the least previous heat exposure. The physician's diagnosis was undoubtedly correct, at least in part.

However, one must also consider the possibility of effects from hydrocarbons, and even more importantly, the potential for combined effects of hydrocarbons and heat stress. At the beginning of the job, the safety representative measured a combustible gas level of less than 2.5% of the LFL for methane. Since most of the hydrocarbons that were probably present have an LFL less than methane, a correction must be made to this reading (see earlier discussion and Table 4.1). If we assume that the LFL for hexane is representative of the total atmosphere and that the reading was 2.5%, then the 2.5% reading must be multiplied by 2.5. This factor (2.5) is not presented in the GasTech manual; it must be obtained from GasTech by telephone or correspondence.

Consequently, the reading of 6.25% of the LFL for hexane would be a volume-to-volume concentration of about 0.007%, since the LFL for hexane is 1.2% v/v. Therefore, less than 700 ppm of the atmosphere were hydrocarbon vapors. This argument is fraught with difficulties involving molar volumes and molecular weights in a hetrogeneous atmosphere, but it can serve as a fair estimate of atmospheric concentration.

Therefore, the hydrocarbon concentration at 1:00 p.m. was probably less than 700 ppm, and it was around 700 ppm at 3:45 p.m., as indicated by the TLV Sniffer, which had been calibrated for hexane. This indicates either that the two instruments read equivalently, or that the concentration had changed in the interim. Not enough information is given to deduce any further.

The detector tubes were sensitive enough to be of use but gave results which are difficult to interpret. The toluene 5/a tube indicated 125 ppm. While this tube is also sensitive to xylene, benzene, and other petroleum hydrocarbons, all of these substances give smaller length of stain changes per unit of concentration and slightly different color changes than for toluene. The blend of color changes can make the tubes difficult to read. The benzene 5/a tube indicated presence of other aromatics, since the precleanse section was completely stained, that is, the 5 ppm indicated was not all benzene. The hexane 100/a tube is equally sensitive to cyclohexane but less sensitive to other petroleum fractions. The hydrocarbons 2 tube is sensitive to some aromatics and some aliphatics. Hence, the detector tube readings are not additive; neither are they of any particular use except perhaps to add credence to the TLV Sniffer reading or to obtain an approximation of the constituents of the atmosphere. One would expect the Sniffer results to be higher than the detector tube readings because it senses all hydrocarbons, even though not equally.

With respect to the respirators worn by the workers, one should err on the side of safety and assume that during and before the accident concentrations may have been higher, perhaps as high as 1000 ppm, the upper limit for cartridges. At an assumed minimum protection factor of 10 for the half-mask respirator, the workers would have been inhaling 100 ppm hydrocarbon vapors. Perhaps causing the exposures to be even higher, even if better fits had been attained, is the breakthrough phenomenon. Nelson and Harder demonstrated that at 50% relative humidity, 22°C, and a moderately heavy breathing rate of 53 Lpm, 1% breakthrough was achieved in about 1-2 hr for a number of aromatics and alkanes at 1,000 ppm. At 104°F (40°C) and higher humidities one would expect 1% breakthrough times to be even shorter, and concentrations in the masks after 2 hr exposure to be higher than 10% of ambient. In summary, concentrations in the masks due to leaks and breakthrough could have been 200 ppm or more after 2 hr exposure.

For many hydrocarbons, 200 ppm exposure is not high. However, for some hydrocarbons (see Table 4.4) this level is at, or is higher than, the TLV. More importantly, exposure to the hydrocarbons and the heat stress probably involved an additive or possibly synergistic effect, not to mention the exacerbation of heat stress by respirators. Most hydrocarbons at high

TABLE 4.4
ACGIH Threshold Limit Values and OSHA Permissible Exposure Limits (ppm) for Some Hydrocarbons Derived from Petroleum

Compound	ACGIH (1986)	OSHA
Butane	800	—
Pentane	600	1000
Hexane	50	500
Heptane	400	500
Octane	300	500
Gasoline	300	—
Benzene	10	10
Toluene	100	200
Xylene	100	100
Cumene	50	50
Ethylbenzene	100	100

concentrations have a narcotic effect on the nervous system and produce the symptoms described by the workers; dizziness, fatigue, and headache. Finally, higher breathing rates caused by heat stress will deliver a larger dose per unit time, all else being equal.

Short-Term Controls

The industrial hygienist made three broad recommendations for work in the separator on the following day: (1) Work should be completed early in the morning and prior to the temperature reaching 85°F. Workers should be allowed 15-min breaks each hour. Younger, acclimatized workers should be used if possible. (2) Workers should wear full-face-piece, supplied-air respirators with continuous flow mode. (3) Someone should stand by outside the separator in the event of an emergency and be able to sound an alarm for assistance, if necessary.

If work were started at 7:00 a.m. and completed by 9:00 a.m., this should assure that the temperature does not exceed 85°F. At this time of morning, relative humidity should be around 85% and declining, yielding a *minimum* wet bulb of 75°F. The sun should be of lesser importance with respect to radiation due to the angle of inclination. Therefore, the WBGT (0.7 × 75°F) + (0.3 × 85°F) should be around 26°C, which complies with the ACGIH recommendation for continuous, moderate work. If it were assumed that there were no wind whatsoever (a good assumption), and therefore, NWB = DB = GT, then WBGT = 85°F or 29°C. At this temperature, light continuous or moderate work with 25–50% rest time per hour would be allowed. Therefore, workers could work at temperatures beyond 85°WBGT, but the industrial hygienist should make more accurate readings; workers should always be acclimatized if ACGIH recommendations are exceeded. The WBGT Index should not be greatly exceeded even if the workers are acclimatized, unless it is known that they have done the work before, under similar conditions, without adverse effects.

Workers should be given 15-min rest breaks each hour and an ample supply of cool water. With regard to acclimatization, while it is true that older workers can be even better suited for heat stress than younger workers, younger workers are generally better utilized for heat stress situations where nothing is known about previous acclimatization.

For several reasons, an argument can be made for air-line-supplied, continuous flow, full-face-piece respirators. The continuous flow will help cool the worker and make breathing easier, alleviating some of the work load. The full-face-piece and continuous flow mode will increase the protection factor to 10,000 if an escape bottle is provided. While this protection factor may seem excessive, it should be realized that without the bottle this configuration is good only in concentrations up to the IDLH level (about 2,000–4,000 for most hydrocarbons). Bottled air should be used or, if an air compressor is used, it should be away from the separator, and one should make certain that the air supplied meets OSHA breathing air standards.

The egress is not restricted, since the workers should be able to climb out of the separator without the aid of the hoist. In the event that workers faint or become incapacitated, the standby worker at the edge of the separator should sound an alarm for assistance. Nevertheless, it may be wise to issue the workers 5-min escape air bottles and properly train them in their use.

Long-Term Controls

The major outcome of the long-term study is to verify the need for more effective respirators. Table 4.2 shows that at least two types of changes in hydrocarbon concentrations are possible in the separator. The first is characterized by day 4, shift 1; day 3 and possibly day 5 of shift 2; and day 6, shift 3. In each case, the concentrations are 2–3 times higher than on other shifts, indicating that some process change is taking place. These data should be of importance to plant engineers. If the high vapor occurrences can be traced to changes in process flows or temperatures or out-of-control situations, corrections could possibly be made, resulting in less product loss, a savings to the company.

In any event, exposures will be two to three times normal for these situations. If we can assume that the analyzed hydrocarbons constitute about 8% of the total fraction (see Table 4.3), then we can estimate the total concentration of hydrocarbons. (Note: We must also assume that the other, heavier hydrocarbons also volatilize in proportion to their solution concentrations, which is not exactly correct, but errs on the safe side.) The day 4, shift 1 total concentration would equal the total concentration of those compounds analyzed multiplied by 12, or about 500 ppm. For day 1, shift 1, the total would be only about 300 ppm.

For future use of these data, correlation with combustible gas-meter readings at the time of study would have been far-sighted. Unfortunately, this was not done. In any event, the air-supplied, full-face respirator should prevent overexposure. However, one other aspect is of importance. NIOSH recommends that workers not enter confined spaces when atmospheres are 10% or greater of the LFL. Although it was argued earlier that the separator is a marginal confined space, nevertheless, the data in Table 4.2 and assumptions made in the previous paragraph indicate that the total concentration of hydrocarbons could be as high as 2800 ppm. (See day 5, shift 3; more will be said about this shift later.) If it is again assumed that the LFL for hexane is a good approximation of the total atmosphere, then 10% of 1.2% is 0.0012 or 1200 ppm. Therefore, in this rare case the concentration is about 30% of the hexane LFL. In light of this finding, a continuous monitoring combustible gas meter with alarm, such as the GasTech model 1641 (there are many others available), should accompany the workers into the separator.

Day 5, shift 3 concentrations are so unusual that further investigation was made into their possible cause. Note that the concentrations of alkanes are about normal, but those of aromatics are at least an order of magnitude higher than normal. This would imply that the primary petroleum refining process was not the cause of this occurrence, but rather some specialized process dealing with aromatic distillates. In this case, the benzene and toluene (BTU) and xylene units, or perhaps the tank farm are possibilities. After looking at production records for the BTU, it was discovered that there had been a spill of stabilized platformate or unseparated benzene, toluene, and xylene on that day. The spill was washed into the sewer. Normal procedure is to vacuum the product from the top of the sewer. However, probably due to the fact that the graveyard shift was on duty, this step was not performed, and the product went to the separator. Consequently, these kind of random, yet highly possible, occurrences can take place in a large refinery and should be expected.

A final consideration is the use of qualitative tests for determining respirator fit. In situations where workers are exposed to consistant, known concentrations, qualitative fit tests provide a level of assurance that workers have some minimal level of protection. However, when air concentrations of a contaminant are variable, it is not known if the protection factor is adequate for all situations. One solution is to have the worker wear a positive-pressure respirator or self-contained breathing apparatus that gives maximum protection. Another approach is to use quantitative fit testing and a program of air sampling that quantifies, as well as possible, all contaminant concentrations. It should also be noted that false-positive qualitative tests results are possible, and these may only be uncovered after a quantitative fit test or after serious harm is done to the worker.

In summary, the recommendations are:

1. Perform work early in the morning or when the temperature is below 85°F and try to use acclimatized workers, or young workers if work histories are uncertain.
2. Use full-face-piece, supplied-air, continuous flow respirators.
3. Use a continuous combustible gas monitor, with alarm set at 10% LFL, in the separator at all times. It should be calibrated with hexane.

Alternative Solutions

As with any industrial hygiene problem, many solutions are possible, and those presented here may not necessarily seem optimal. (This is particularly true in light of the probable limited familiarity on the part of the reader with the nature of the refining process.)

In looking at the surveys done by the industrial hygienist immediately following the accident, it seems probable that he suspected hydrocarbon overexposure more so than heat overexposure. For whatever reason, he did not measure WBGT that afternoon. Instead, he chose to estimate this value. Perhaps he did not have a WBGT apparatus. Certainly in the southern United States, one should not overlook the possibility of heat exhaustion.

Rather than using the detector tubes and the TLV Sniffer to measure hydrocarbon concentration on the day of the incident, the industrial hygienist could have ordered the respirator changes without corroborative data. In fact, the industrial hygienist might have elected to use a cool air vortex and supplied-air respirator in combination. This would have substantially reduced risks posed by heat stress and hydrocarbon exposure.

In a highly complex plant, with several thousand employees, and only one industrial hygienist, an error concerning respirator usage and maintenance work may not be uncommon. This error can be partially due to lack of knowledge or training or even laziness on the part of the maintenance foreman, industrial hygienist, or safety representative. Most likely it is due to failure to document, or for management to enforce, standard safety operating procedures, including a respirator policy. Documentation and enforcement is a must.

Another potential source of data on the day of the incident were analyses performed as part of compliance with regulations under the Clean Water Act. These require that more than 100 priority pollutants be monitored in waste water. While these data surely would not have provided a quantitative assessment of air concentrations, it is possible that data collected frequently enough may have given the industrial hygienist a clue as to how often and to what extent pollutant levels change in the separator.

With respect to the long-term study, obviously many designs are possible. In this case, the industrial hygienist expected some rather infrequent, high contaminant concentrations due to process problems and spills. Had he not expected these and sampled for only a few days, he would have confirmed his expectations and missed a potential problem. As it turns out, he may have been fortunate to detect the problem by doing the "limited" sampling that he did. For example, if an unusual event only has only a 5% chance of occurring in any given time period (say 2 hr), and one randomly samples 14% of the time periods, as was done here (48 2-hr samples out of 336 2-hr periods over 28 days), then the chance of observing an unusual event is less than 1%.

One way to avoid this problem is to use a direct-reading instrument on a 24-hr basis. However, maintenance and time spent on such a project can be quite high. Also, the instrument may not be appropriate for differentiating a large number of different compounds, as is gas chromatographic analysis. For example, any type of nonchromatographic instrument with a photo- or flame-ionization detector will yield total hydrocarbon results only. Depending upon the purpose of the study, this may or may not be satisfactory. One should always have a purpose before collecting any data. In this case, it appears that the industrial hygienist wanted to track specific organics, so that he might relate high concentrations to specific refinery units.

REFERENCES

American Conference of Governmental Industrial Hygienists. *Threshold Limit Values for 1986/1987*, ACGIH, Cincinnati, OH, 1986.

American Conference of Governmental Industrial Hygienists. *Industrial Ventilation Manual*, 16th Ed., ACGIH, Cincinnati, OH, 1980.

American Petroleum Institute. *Manual on Disposal of Refinery Wastes*: Vol. on *Liquid Wastes*, API, Washington, DC, 1969.

Drägerwerk Ag Lubeck. *Detector Tube Handbook*, Dräger: Pittsburgh, 1983.

Environmental Protection Agency. "Petroleum refining industry," in *Industrial Process Profiles for Environmental Use*, EPA-600/2-77-023C. Research Triangle Park, NC, 1977.

GasTech, *Model 1641 Combination H_2S/Combustibles/Oxygen Indicator and Alarm*, 1641-012377-0, Mountain View, CA.

Jones, H. R., *Pollution Control in the Petroleum Industry*, Noyes Data Corporation, Park Ridge, NJ, 1973.

Mine Safety Appliance Co. *Charcoal Sample Selection Tubes Instruction Sheet*, Pittsburgh, Part No. 459004.

Nelson, G.O., and Harder, C.A. Respirator cartridge efficiency studies. VI. effect of concentration, *Am. Ind. Hyg. Assoc. J.* 37:205, 1976.

5

HEAT TREATING IN THE METALS INDUSTRY

R. KENT OESTENSTAD

5.1. PROBLEM RECOGNITION

Process Description

In the equipment manufacturing industry, many metal alloy parts must be heat treated to improve their strength, hardness, durability, and resistance to heat, impact, and corrosion. Surfaces are hardened by diffusing carbon or nitrogen into the metal to a given depth to achieve the desired hardness properties. A common hardening method is called case hardening. In this process, the piece is usually heated to 870–980°C (1600 - 1800°F) in an atmosphere containing high concentrations of carbon monoxide, which serves as the source of the diffused carbon.

There are several classes of gas carburizing operations, with the gas-generation technique being the most important variable. Typically, the composition of gas carburizing atmospheres is 40% nitrogen, 40% hydrogen, and 20% carbon monoxide. The carbon monoxide percentage must be regulated closely to assure that a sufficient amount of carbon is available to achieve the desired metal properties without scaling or disfiguring the steel (Burgess, 1981).

Occupational health hazards resulting from carburizing operations include exposures to the furnace gas mixture and heat-stress conditions. Emissions from carburizing furnaces are controlled by minimizing leaks in equipment, providing local exhaust ventilation at furnace vents, and by using flame curtains at furnace doors to control escaping gases (i.e., provide for further combustion of gases to yield less toxic compounds).

However, even under controlled conditions, workplace exposures of greater than 100 ppm of carbon monoxide can occur in gas carburizing operations. In addition, if parts are not cleaned before entering the process, volatile materials may be driven off by the heat and result in worker exposures to toxic contaminants. For example, significant exposures to smoke and organic degredation products such as benzo[*a*]pyrene can result from heat treating parts that are contaminated with cutting oil.

Nature of the Problem

In some carburizing processes it is desirable to harden only portions of a part while leaving the remainder untreated to allow for additional machining. In these situations, carbon diffusion is prevented on the untreated portion by application of a noncombustible, inert coating called a protectant. The uniform composition and thickness of this coating is critical to assure proper heat treating. Also, the material must be easily removed after carburizing is completed. There are several commercial formulations available to meet these specifications. Depending on the coating properties desired, some of these coating products use organic solvents as vehicles, while others are water-based. The coatings are usually applied by brush or dipping.

The industrial equipment manufacturing facility in this study includes a large heat-treating department in which as many as 15 treatment processes are conducted simultaneously. The department is housed in a 80 ft × 300 ft building that has louvered outside walls along the length of the building to allow for maximum natural dilution ventilation. The louvers are opened only during fair weather, and the amount and quality of dilution air is affected by meteorlogical conditions such as wind speed and direction, humidity, and temperature.

The plant industrial hygienist received a request for an exposure evaluation from the plant medical department near the end of the first shift (3:30 p.m.) on a day in late June. An employee who was applying a carburizing protectant compound to transmission gears had reported to the medical department near the end of his shift. The worker had complained of having a severe headache, blurred vision, dizziness, and drowsiness.

The industrial hygienist immediately went to interview the employee. In this interview it was learned that the employee had been doing the same work for a week, and the symptoms had developed over the past 2 days. After talking with the employee, the industrial hygienist contacted the first-shift supervisor. The supervisor confirmed that the operation had been running for only 1 week and was developed for parts of a modified transmission, which recently went into production. Because of high initial production schedules for this component, the part was being run on all three work shifts for an indefinite period of time. The supervisor stated that she had noticed a "paint" odor when in the area where the operation was conducted, but that neither the second- nor third-shift operators had complained of any problems with the new procedure. It was the industrial hygienist's responsibility to ascertain the nature and extent of the exposure and to recommend control procedures if an overexposure to an occupational health hazard was identified.

After interviewing the affected employee and his supervisor, the industrial hygienist contacted the chemical engineering department to obtain the name and identification number of the coating compound being used. This information was used to retrieve the product's material safety data sheet (MSDS) (Figure 5.1) from the company's hazardous materials information file.

With this information, the industrial hygienist went to the heat-treating department to observe the operation during the second shift, to become familiar with the layout of the work area, and to make preliminary measurements of airborne toxic substances. The job was observed to consist of three steps. The first step involved loading 36 gears into furnace racks that were then put into a continuous, preheat furnace. In the next step the parts from the preheat furnace were moved on a roller conveyor to a work station where the carburizing protectant was painted on a portion of the gear, and they were then placed into a carburizing furnace. The final step involved the removal of the part racks from the carburizing furnace and their transfer to an adjacent "laydown" area for cooling. Each part weighed about 5 lb; therefore, powered hoists were used to lift the loaded racks.

The industrial hygienist observed the affected employee using a small brush to apply the protectorant from an open 1-gal can. It was necessary for him to stir the mixture frequently to keep the solids suspended in the solvent. To maintain proper consistency of the protectorant, small amounts of solvent were periodically transferred in a small open can from a 5-gal container to the protectorant can. The MSDS for this solvent was obtained (Figure 5.2) and reviewed. The only personal protec-

Material Safety Data Sheet

Required under USDL Safety and Health Regulations for Shipyard Employment (29 CFR 1915)

OMB No 1218-0074
Expiration Date 05/31/86

Section I

Manufacturer's Name		Emergency Telephone Number
Metal Coat Products		
Address (Number Street City State and ZIP Code)	Chemical Name and Synonyms	
Postfach 309, 737 West Germany	Trade Name and Synonyms	CONDURSAL 0090
	Chemical Family	Formula
Compound for protecting parts during gas carburizing		Proprietary

Section II - Hazardous Ingredients

Paints, Preservatives, and Solvents	%	TLV (Units)	Alloys and Metallic Coatings	%	TLV (Units)
Pigments Boron Oxides	50		Base Metal		
Catalyst			Alloys		
Vehicle Resin	25		Metallic Coatings		
Solvents Aromatic Hydrocarbon	25		Filler Metal Plus Coating or Core Flux		
Additives			Others		
Others					

Hazardous Mixtures of Other Liquids, Solids or Gases

	%	TLV (Units)

Section III - Physical Data

Boiling Point (°F) from solvents:	284 °F	Specific Gravity (H_2O=1)	1.3
Vapor Pressure (mm Hg) at 68°F	6 torr	Percent Volatile by Volume (%)	30
Vapor Density (AIR=1)		Evaporation Rate (Ether =1)	13.5

Solubility in Water

No

Appearance and Odor

Section IV - Fire and Explosion Hazard Data

Flash Point (Method Used)	Flammable Limits	Lel	Uel
from solvents: 74°F		1.0	7.6

Extinguishing Media

CO_2

Special Fire Fighting Procedures

Use no water

Unusual Fire and Explosion Hazards

None

Page 1 (Continued on Reverse Side)

Form OSHA-20 (Rev 3/84)

FIGURE 5.1a. Material safety data sheet for protectorant.

Section V - Health Hazard Data

Threshold Limit Value

200 ppm

Effects of Overexposure

Severe eye irritation, drying and possible burning of skin. Excessive inhalation causes headache, dizzyness, nausea, increasing signs of anesthesia.

Emergency First Aid Procedures

Eye contact: flush with water. Skin contact: wash with mild soap and apply skin cream.

Inhalation: if illness occurs, remove patient to fresh air, keep quiet and warm.

Ingestion: induce vomiting only at physician's recommendation.

Section VI - Reactivity Data

Stability	Unstable		Conditions to Avoid Heat, sparks, open flame and fire
	Stable	X	

Incompatability (Materials to Avoid)

Hazardous Decomposition Products

Hazardous Polymerization	May Occur		Conditions to Avoid
	Will Not Occur	X	

Section VII - Spill or Leak Procedures

Steps to be Taken in Case Material is Released or Spilled

Dry and remove

Waste Disposal Method

Section VIII - Special Protection Information

Respiratory Protection (Specify Type)

Ventilation	Local Exhaust X	Special
	Mechanical (General)	Other

Protective Gloves	Eye Protection Yes

Other Protective Equipment

Section IX - Special Precautions

Precautions to be Taken in Handling and Storing

Cool and closed

Other Precautions

Page 2

Form OSHA-20 (Rev 3/84)

FIGURE 5.1*b*. Material safety data sheet continued.

Material Safety Data Sheet

Required under USDL Safety and Health Regulations for Shipyard Employment (29 CFR 1915)

OMB No 1218 0074
Expiration Date 05/31/86

Section I

Manufacturer's Name		Emergency Telephone Number
Metal Coat Products		
Address (Number, Street, City, State and ZIP Code) Postfach 309, 737 West Germany	Chemical Name and Synonyms	
	Trade Name and Synonyms	CONDURSAL SOLVENT
	Chemical Family	Aromatic Hydrocarbon

Formula

Section II - Hazardous Ingredients

Paints, Preservatives, and Solvents	%	TLV (Units)	Alloys and Metallic Coatings	%	TLV (Units)
Pigments			Base Metal		
Catalyst			Alloys		
Vehicle			Metallic Coatings		
Solvents Aromatic Hydrocarbon	100		Filler Metal Plus Coating or Core Flux		
Additives			Others		
Others					

Hazardous Mixtures of Other Liquids, Solids or Gases

	%	TLV (Units)

Section III - Physical Data

Boiling Point (°F)	231°F	Specific Gravity (H_2O=1)	0.86
Vapor Pressure (mm Hg) at 68°F	22	Percent Volatile by Volume (%)	100
Vapor Density (AIR=1)	3.14	Evaporation Rate Butyl Acetate =1)	2.24

Solubility in Water

Appearance and Odor

Section IV - Fire and Explosion Hazard Data

Flash Point (Method Used)	Flammable Limits	Lel	Uel
40°F		1.3	7.1

Extinguishing Media

CO_2

Special Fire Fighting Procedures

Use no water

Unusual Fire and Explosion Hazards

None

Page 1 (Continued on Reverse Side)

Form OSHA-20 (Rev 3/84)

FIGURE 5.2a. Material safety data sheet for protectorant solvent.

Section V - Health Hazard Data

Threshold Limit Value

For solvents only: 200 ppm

Effects of Overexposure

Severe eye irritation, drying and possible burning of skin, excessive inhalation, headache, dizzyness, nausea, increasing signs of anesthesia.

Emergency First Aid Procedures

Eye contact: flush with water. Skin contact: wash with mild soap and apply skin cream. Inhalation: if illness occurs, remove patient to fresh air, keep quiet and warm. Ingestion: induce vomiting only at physician's recommendation.

Section VI - Reactivity Data

Stability	Unstable		Conditions to Avoid: Heat, sparks, open flame and fire.
	Stable X		

Incompatability (Materials to Avoid)

Hazardous Decomposition Products

Hazardous Polymerization	May Occur		Conditions to Avoid
	Will Not Occur	X	

Section VII - Spill or Leak Procedures

Steps to be Taken in Case Material is Released or Spilled

Dry and remove

Waste Disposal Method

Section VIII - Special Protection Information

Respiratory Protection (Specify Type)

Ventilation	Local Exhaust X	Special
	Mechanical (General)	Other

Protective Gloves	Eye Protection Yes

Other Protective Equipment

Section IX - Special Precautions

Precautions to be Taken in Handling and Storing

Cool and closed

Other Precautions

Page 2

Form OSHA-20 (Rev 3/84)

FIGURE 5.2*b*. Material safety data sheet continued.

tive equipment used by the employee were cotton gloves. The operation was continuous, and about 20 racks were completed in an 8-hr work shift. The employee spent approximately 2 hr loading racks into the preheat furnace; 4 hr removing racks from the preheat furnace, applying carburizing protectant, and loading the coated parts into the carburizing furnace; and about 2 hr unloading the racks from the carburizing furnace. Figure 5.3 illustrates the layout of the work area.

Based on the symptoms experienced by the first-shift operator, the information on the MSDS and observation of these

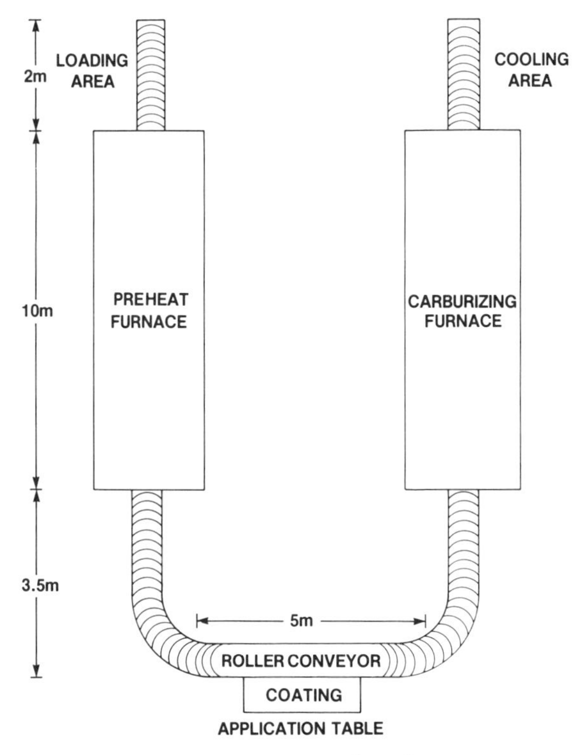

FIGURE 5.3. Layout of work area.

work practices, the industrial hygienist required that respiratory protection be worn by the operators until the extent of the exposure was identified.

Questions

1. What environmental contaminants and/or conditions would you measure?
2. What sampling methods would you use to measure the exposures identified in Question 1?
3. Are the first-shift employee's symptoms consistent with the toxicity of the known contaminants? Should exposure to these materials be considered as being additive? If so, what standards or professional guidelines would be applicable?
4. In addition to environmental measurements, what other actions should be taken? What safety hazards might be present?
5. Does the information on the material safety data sheets appear to be adequate? If not, how might you obtain additional information to better define the exposure problem?
6. Does the decision to require respiratory protection seem prudent in consideration of the information available at the time? Discuss.
7. What type of respirator should be used? What are the criteria for selection of respirators?

5.2. EVALUATION RESULTS

Preliminary Findings

Preliminary measurements for gases and vapors were made with Drager indicator tubes. Results of these "grab-sample" measurements are shown in Table 5.1.

During the first shift on the following day, personal samples were collected from the employee who had complained of health problems. A charcoal tube sample for organic solvents was collected using a DuPont P-200 low volume sampling pump calibrated at 100 cm^3/min. Also, to estimate the time-weighted exposure to carbon monoxide, a Drager long-term indicator tube was coupled with a DuPont P-125 low-volume sampling pump calibrated at 15 cm^3/min and was worn by the worker for the full shift (Lioy and Lioy, 1983). At the end of the shift, this tube indicated an average carbon monoxide exposure of 10 ppm. In addition, measurements for heat-stress conditions were made at the three work stations using a WIBGET heat-stress monitor, a sling psychrometer, and an Alnor Thermoanemometer. Results of the heat-stress measurements are given in Table 5.2.

To obtain more specific information on the composition of the coating product, the coating supplier was contacted by telephone. The supplier's technical representative stated that the product was formulated in the Federal Republic of Germany and that the aromatic hydrocarbon listed on the MSDS was toluene. Based on this information, the charcoal tube and a bulk sample of the coating material were sent to a consulting analytical laboratory with a request for analysis for toluene by gas chromatography (Eller and Crable, 1979). The laboratory report of analysis of the charcoal tube sample for toluene was received the following week. The air sample indicated an 8-hr, time-weighted average exposure of 65 ppm toluene.

In-Depth Studies

Even with the limited amount of data available, the industrial hygienist thought that the symptom reported by the employee were inconsistent with the measured exposure levels. Because of these questions, a gas chromatography/mass spectrographic analysis of the coating bulk sample was requested before continuing with any additional exposure measurements

TABLE 5.1
Preliminary Gas and Vapor Measurements

	Air Concentration (ppm)		
Contaminant	Furnace Loading	Coating Area	Furnace Unloading
Carbon monoxide (5/c)	10	5	10–15
Toluene[a] (25/a)	ND[b]	100–150	ND[b]

[a]Toluene was selected as a representative aromatic solvent, since no specific compound was identified from the material safety data sheet.
[b]ND, none detected.

TABLE 5.2
Heat-Stress Measurements

Parameter	Furnace Loading	Coating Area	Furnace Unloading
Natural wet bulb temp (°C)	24	24	25
Globe temp. (°C)	33	28	34
Dry bulb temp. (°C)	28	28	29
Forced wet bulb temp. (°C)	22	22	22
Air velocity (m/min)	65	65	97
Estimated work load (cal/hr)	300	300	300

(Eller and Crable, 1979). The report of that analysis showed that the coating compound solvent consisted of 80% toluene and 20% chlorotoluene. Upon receipt of this information, additional breathing-zone samples using charcoal tubes were collected on 2 consecutive workdays on both the first and second shifts. The exposed tubes were sent to the analytical laboratory for gas chromatographic analysis for toluene and chlorotoluene. Also, long-term Drager indicator tube samples were repeated to measure carbon monoxide exposures. Results of these measurements are summarized in Table 5.3.

TABLE 5.3
8-Hr, Time-Weighted Breathing Zone Exposure to Gases and Vapors

	Concentrations (ppm)		
Sample	Toluene	Chlorotoluene	Carbon Monoxide
Day 1			
1st Shift	60	10	10
2nd Shift	50	8	10
Day 2			
1st Shift	70	12	15
2nd Shift	55	8	10

Measurements were also made for phosgene during these four work shifts using Drager Phosgene 0.05/a detector tubes (sensitivity of 0.05 ppm). No detectable concentrations were found.

Questions

1. Discuss the presence of a large percentage of unreported chlorotoluene in the coating compound? How does the toxicity of chlorotoluene compare to that of toluene?
2. Why were samples collected for phosgene?
3. Calculate the 95% upper and lower control limits (UCL and LCL) for the exposure data in Table 5.3 using the coefficients of variation (CV's) provided in the appropriate NIOSH Analytical Method or by the manufacturer of the indicator tubes.
4. Even with the additional exposure to chlorotoluene, is there an overexposure to the mixture of contaminants? Is the amount of exposure data sufficient to make this

decision? How should the industrial hygienist deal with exposures that are at or near the allowable limits and limited exposure data?

5. Should the exposure to the identified contaminants be considered as being additive? If so, how would the exposure to the mixture of contaminants be calculated?
6. What controls might be applied to reduce or eliminate these exposures? Which would be most applicable in this situation? Are the heat-stress conditions significant?
7. How could this exposure situation have been avoided? At what point in the production process would the involvement of the industrial hygienist been most beneficial?

5.3. CONTROL STRATEGIES

Discussion

Upon notification of adverse health symptoms resulting from a possible occupational exposure, the initial step of an evaluation is to determine what exposures might exist and what factors might affect those exposures. Once this information is determined, a sampling strategy can be formulated to evaluate the exposures. This initial assessment requires a knowledge of the process and materials being used. In heat-treating operations, heat-stress and carbon monoxide exposures are usually involved because of the high levels of conductive and radiant heat and the extremely high concentrations of carbon monoxide used in carburizing operations. Additional exposures may also result from special operations or from heating contaminated parts. In this case study, solvent exposures resulted from the application of a carburizing protectant compound. If the industrial hygienist is familiar with the operation and has access to material composition information, the initial assessment can be made immediately and preliminary evaluations obtained within a short time.

Ideally, once an occupational health problem is identified, it would be desirable to avoid any additional exposures until their nature and extent could be determined. In some production situations this can be accomplished by temporarily suspending the operation; if not, the use of personal protective equipment may be necessary. Because of high production demands, the latter was the situation in this case. Employees on three work shifts were exposed continuously from the initiation of the operation until respirators were required after the industrial hygienist had observed the operation and made a preliminary assessment of exposure potential.

Exposure to toluene, chlorotoluene, and carbon monoxide can be considered to affect the same organ systems; therefore their combined effect should be given primary consideration. Both toluene and chlorotoluene are central nervous system depressants. Carbon monoxide indirectly affects the central nervous system by limiting the amount of oxygen transported in the blood. The TLV for mixtures is calculated from the equation in Appendix C of the *ACGIH Threshold Limit Values*. Applying that formula to the three exposures presented in Table 5.3 gives the values in Table 5.4.

These calculations suggest that the first-shift operator is overexposed to the mixture, and the second-shift operator is exposed to levels that approach the combined TLV. Considering the accuracy of sampling methods and variations in individual susceptibility, these exposures, in addition to the heat-stress conditions, could account for the symptoms experienced by the first- and second-shift operators. Also, it would be reasonable to expect that the contaminant concentrations would increase when the natural dilution ventilation is reduced in the heat-treating building as a result of the wall louvers being closed during inclement weather. The conditions would justify recommendations for permanent controls beyond the use of respirators.

Heat-stress conditions do not directly affect the central nervous system as do the other contaminants in this case. However, they do limit the body's ability to metabolize and excrete

TABLE 5.4
Consideration of Combined Exposures

Sampling Period	Combined Exposure Ratio
Day 1	
1st Shift	1.00
2nd Shift	0.86
Day 2	
1st Shift	1.24
2nd Shift	0.91

absorbed toxic materials, since the flow of blood is shunted to the skin and away from vital organs or metabolic sites (Mutchler, 1978). Therefore, exposure to heat can potentiate the effects of other contaminants and should be included in the hazard evaluation. Initial measurement of heat-stress conditions revealed that the first-shift operator was exposed to conditions that were at or near the limits recommended by the ACGIH for continuous, moderate work loads. From the data in Table 5.2, the calculated average WBGT for the entire first shift is 26.5°C, as compared to the recommended TLV of 26.7°C for moderate work loads. Radiant heat was a significant factor at the carburizing loading and unloading stations. Other factors that appeared to affect heat-stress levels were metabolic rate and prevailing weather conditions. Therefore, it could be expected that heat stress would increase as production levels increased, and as temperatures and humidity increased during the months of July and August.

Information provided on MSDS is often incomplete or, as in this case, inaccurate. It is advisable to exercise caution when interpreting information on these forms, even if they are filled out in great detail. Often it is necessary to contact the distributor or manufacturer to obtain additional information or to confirm data already provided. In this example, it was necessary to go one step further and have an expensive analysis performed on the material to confirm its composition. The presence of such a large percentage of unreported solvent would suggest either a gross lapse of production quality control or intentional omission by the manufacturer.

When reviewing a MSDS, the industrial hygienist should watch for inconsistencies between information provided in the various sections. A common example is that the manufacturer will not identify any hazardous ingredients in Section II but will list symptoms of overexposure in Section V and specify special protection information in Section VII. Often, the manufacturer omits hazardous ingredients because they are not regulated by OSHA Standards. Many times, the industrial hygienist must make his/her own hazard evaluation, based on the composition of the product and corresponding toxicological data that is available in the literature.

When making control recommendations, it is important to be sensitive to production methods and materials considerations. In this case, the gear parts being treated had been designated as "high priority," and any recommendation that would limit production output would have been difficult to implement. Therefore, any change in the job method that would require a reduction in productivity to reduce exposure levels would require substantial justification. Changes in materials (coating compound) might also affect productivity. In addition, metallurgical requirements must also be taken into account when recommending substitution of materials. An engineering control applicable in this situation would be local exhaust ventilation at the work station where the coating compound is applied to the gears. Application of this control would include significant capital and operational costs; therefore a well-developed recommendation with good supporting information would be important.

Short-Term Controls

In this case, implementation of controls was accomplished sequentially. Initially, personal protective equipment was required. This temporary control was instituted before the exposure evaluations had been completed and included the use of half-mask, air-purifying respirators with organic vapor cartridges. This type of respirator was selected through use of the protocol recommended in ANSI Z88.2 *Practices for Respiratory Protection*. In addition, a complete respiratory protection control program was implemented in accordance with OSHA standards (CFR 1910.134). Half-mask respirators were thought to be adequate protection because of the good warning properties of toluene (i.e.: characteristic odor, and early symptoms of overexposure). Preliminary measurements (Drager and charcoal tubes) indicated that peak exposure levels should not exceed 200 ppm (Birkner, 1980).

In addition to the respirators, Viton gloves were required on a permanent basis to prevent irritation of the skin and absorption of the solvent through the skin. Although Viton is an expensive glove, it provides the best protection for toluene and chlorinated aromatics, as stated in glove selection charts based on permeation properties and provided by the manufacturer. In addition, the open cans of solvent were replaced with safety cans.

Long-Term Controls

The permanent control recommendation involved the use of local exhaust ventilation at the work station where the coating material was applied. A side-draft slot hood similar to that shown in Figure 5.4 was designed and installed immediately behind the roller conveyor between preheat and carburizing furnaces (ACGIH, 1984). The hood was made long enough to provide ventilation at the point where the coating material was

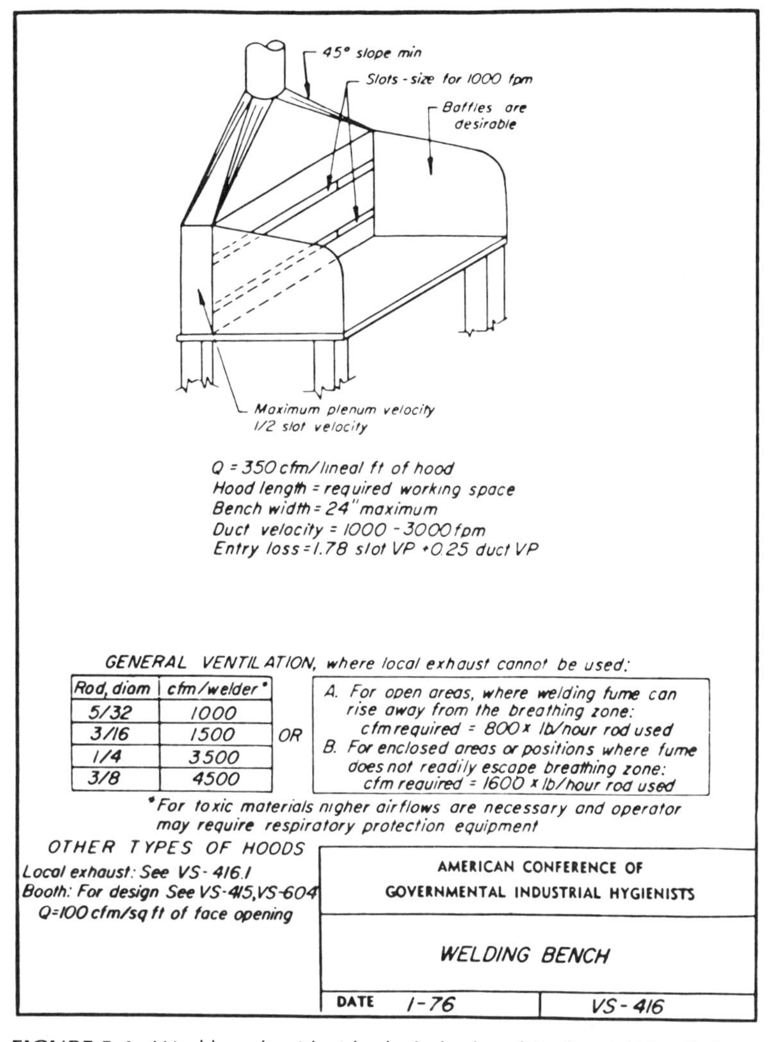

FIGURE 5.4. Workbench with side-draft slot hood (*Industrial Ventilation*, 17th Ed., ACGIH).

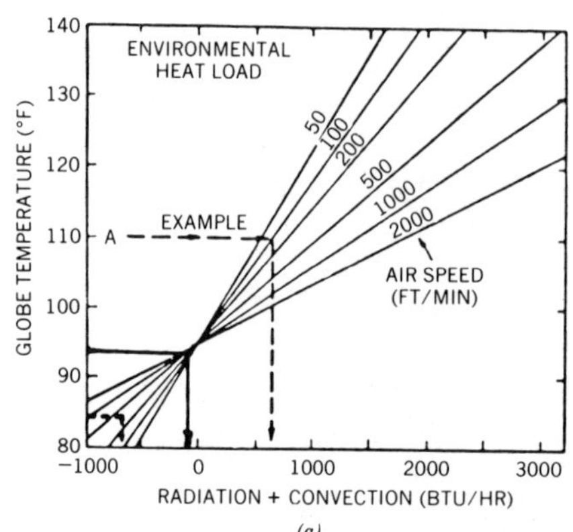
ENVIRONMENTAL HEAT LOAD
EXAMPLE
A
50
100
200
500
1000
2000
AIR SPEED (FT/MIN)
GLOBE TEMPERATURE (°F)
140
130
120
110
100
90
80
−1000
0
1000
2000
3000
RADIATION + CONVECTION (BTU/HR)

(a)

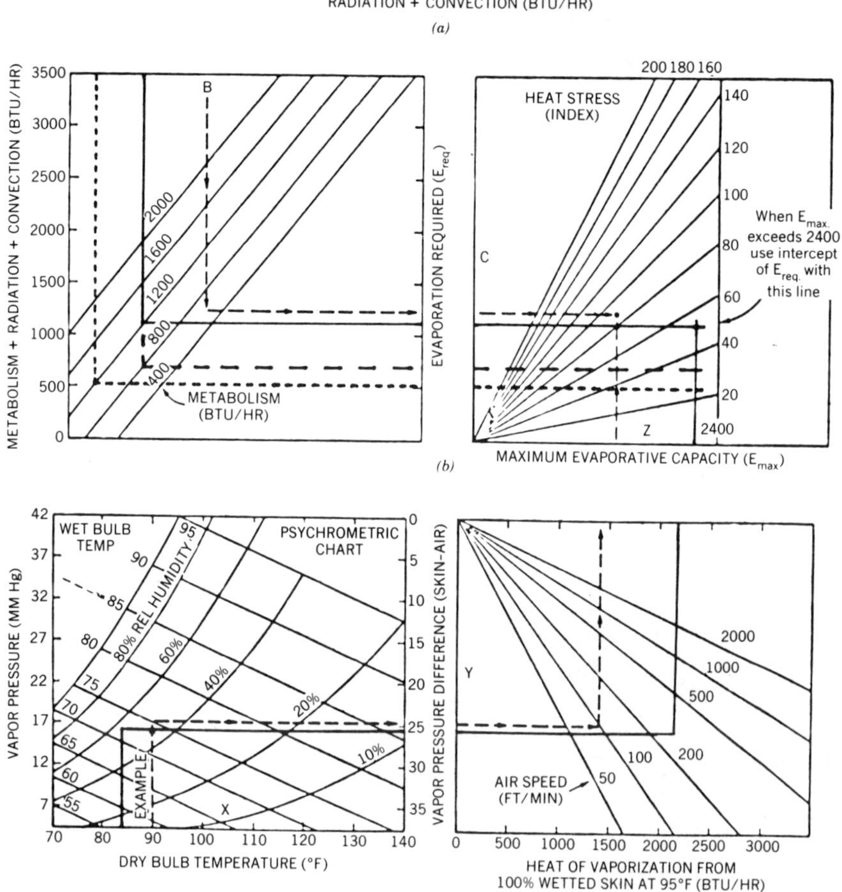
METABOLISM + RADIATION + CONVECTION (BTU/HR)
3500
3000
2500
2000
1500
1000
500
0
B
2000
1600
1200
800
400
METABOLISM (BTU/HR)
HEAT STRESS (INDEX)
200 180 160
140
120
100
80
60
40
20
2400
C
Z
EVAPORATION REQUIRED (E_req)
When E_max. exceeds 2400 use intercept of E_req. with this line
MAXIMUM EVAPORATIVE CAPACITY (E_max)
(b)
WET BULB TEMP
PSYCHROMETRIC CHART
95
90
85
80
75
70
65
60
55
80% REL HUMIDITY
60%
40%
20%
10%
EXAMPLE
X
VAPOR PRESSURE (MM Hg)
42
37
32
27
22
17
12
7
70
80
90
100
110
120
130
140
DRY BULB TEMPERATURE (°F)
VAPOR PRESSURE DIFFERENCE (SKIN-AIR)
0
5
10
15
20
25
30
35
Y
2000
1000
500
200
100
50
AIR SPEED (FT/MIN)
0
500
1000
1500
2000
2500
3000
HEAT OF VAPORIZATION FROM 100% WETTED SKIN AT 95°F (BTU/HR)

(c)

applied, as well as an area where the solvent could evaporate from the coated gears. The initial capital cost of the ventilation system was approximately $7000 for the hood, duct, and fan, plus about $3000 for installation.

Following installation and start-up of the ventilation system, exposures to toluene and chlorotoluene were reevaluated on 3 consecutive days. These measurements found no detectable concentrations of the solvents. As a result, the use of respiratory protection by the operators was discontinued. Four months had passed from the time of the first report of exposure until the effectiveness of the new ventilation system was documented.

While the WBGT Index is a good indicator of heat-stress conditions, it is not as well suited to predicting the efficacy of possible control measures. The Heat Stress Index (HSI) is more useful in estimating the impact of changes in heat-stress factors. Determination of the HSI results in more knowledge about the environment, and use of the HSI nomogram (Figure 5.5) will illustrate the relative contribution of each heat-stress

---

FIGURE 5.5. Heat Stress Index nomogram. Enter A. Drop vertical line from intercept of globe temperature with air speed; thus get combined radiation and convection heat load. Extend vertical line to Enter B. At intercept with metabolism, draw horizontal line; obtain total heat load in terms of evaporation required for heat balance E_{req}. Extend horizontal line to Enter C. Enter X. Draw horizontal line from intercept of db and wb temperature: obtain vapor pressure gradient between saturated skin at 95°F and ambient air. Extend line to Enter Y. At intercept with air speed draw vertical line; obtain maximum evaporation from wet skin at 95°F (E_{max}). Extend line to Enter Z. Move to intercept with horizontal lines from C. (If E_{max} exceeds 2400, enter Z at 2400.). Read Heat Stress Index value.

Conditions at the furnace unloading station are plotted to yield a HSI value of about 50. Change in HSI by reducing globe temperature only is plotted by the dotted line to yield a value of about 25. Change in HSI by reducing metabolic rate only is plotted by the dashed line to yield a value of about 30.

factor. The effect of possible controls on these factors can be illustrated on the nomogram prior to their application (Olishifski, 1979).

An example of how to use the HSI nomogram is presented in Figure 5.5. Values for measurements made at the furnace unloading station are plotted by the solid line in Figure 5.5 to yield a HSI value of about 50. By studying the nomogram, it appears that the HSI could be reduced by reducing globe temperature (radiant heat) to the dry-bulb temperature or by reducing the metabolic rate. The effect of reduction of globe temperature only is shown by the dotted line, and the effect in reduction of metabolic rate only is shown by the dashed line. It would appear that shielding to reduce radiant heat from the furnace might be the most effective control of heat stress in this situation. Reduction of metabolic rate would require a decrease in production or assigning another employee to the job to allow for increased rest periods.

As a final step, the plant industrial hygienist and physician met with the production manager to discuss the problems introduced by this change in the work process. The production manager was made aware of the operator lost time and decreased effectiveness associated with the situation and was advised that involvement of the industrial hygienist early on in the planning and design phases of the new procedure could have prevented the problems from developing.

The initial recommendation for permanent control was to evaluate the use of alternative coating compounds containing less toxic solvents. The chemical engineering department found that there were three other coating compounds available that might have the needed materials properties. Of these, two were water-based products and one used trichloroethane as a solvent. The trichloroethane-based product was eliminated because of the presence of open flames in the heat-treating department, which might cause decomposition of trichloroethane to hydrogen chloride and/or phosgene. Samples of the two water-based products were obtained and evaluated for possible use. This evaluation revealed that one product did not provide the required metallurgical properties and the other could not be easily removed from the gears after carburizing. Therefore the "substitution" control strategy was determined to be unfeasible.

REFERENCES

American Conference of Governmental Hygienists. *Industrial Ventilation*, 18th ed., Cincinnati, OH, 1984.

Birkner, L.R. *Respiratory Protection A Manual and Guideline*, AIHA, Akron, OH, 1980.

Burgess, W.A. *Recognition of Health Hazards in Industry*, Wiley, New York, 1981.

Eller, P.M., and J.V. Crable. "Analytical measurements," in *Patty's Industrial Hygiene and Toxicology*, Vol. III, 1st ed., L.J. Cralley and L.V. Cralley (Eds), Wiley, New York, 1979.

Lioy, P.J., and M.Y. Lioy. *Air Sampling Instruments for the Evaluation of Atmospheric Contaminants*, 6th ed. ACGIH, Cincinnati, OH, 1983.

Mutchler, J.E. "Heat stress: Its effects, measurement, and control," in *Patty's Industrial Hygiene and Toxicology*, 1st ed. G.D. Clayton and F.E. Clayton (Eds.), Wiley, New York, 1978.

Olishifski, J.B. *Fundamentals of Industrial Hygiene*, 2nd ed. Chicago, National Safety Council, 1979.

6

CARCINOGEN EXPOSURE IN A RESEARCH LABORATORY

MAX L. RICHARD

6.1. PROBLEM RECOGNITION

Process Description

The setting is a cancer research laboratory in a major university medical center where the focus has recently turned to the testing of chemicals for chronic health effects, especially their carcinogenic potential. The laboratory is involved in studies of benzidine-based dyes (BBD). Research activities include determination of physical and chemical properties of the dyes and rodent toxicity. Chemical purity of test compounds is determined, "doses" are prepared by mixing dyes with fodder, animal wastes are analyzed for metabolites and parent compounds, and pathological examinations of tissues and organs for tumors are performed. To further define metabolic pathways, some research protocols involve the use of radioactively labeled chemicals. Where studies involve radioactive substances, the samples are prepared and biological waste products, tissues, and organs are analyzed for content of radioactive materials in a radioisotope laboratory.

The laboratory facility has approximately 3000 ft^2 and is located on the sixth floor of a seven-story research building. Thirteen staff members work in the laboratory, and other professionals, from outside the laboratory, such as pathologists and biochemists, occasionally work in the facility. The staff consists of six women, average age 33, and seven men, average age 38. Included on the staff are eight professionals: two biochemists, one pathologist, one geneticist, and four laboratory technicians. Additional staff include two secretaries, two animal attendants, and a custodian.

The laboratory is divided into an administrative support area located at one end and the actual working laboratory or restricted area at the other end (see Figure 6.1). Laboratory staff enter the restricted area only through the men's or women's shower room, where they change from street clothes to laboratory protective wear. The protective wear includes disposable latex gloves, one-piece overalls, shoe covers, and head covering. When staff leave the restricted area for any reason, they are required to shower and to change back into their street clothes.

There are five separate laboratories within the restricted area. The functions of each of the laboratories are chemical analysis; feed blending; radioactive tracer and metabolic studies; pathology; and tissue staining, frozen section, and electron microscopy work.

The laboratories are constructed of gypsum wallboard finished with a high-gloss enamel and have a suspended ceiling system throughout. The heating, ventilation, and air-conditioning system (HVAC) is a 100% make-up air system. Supply

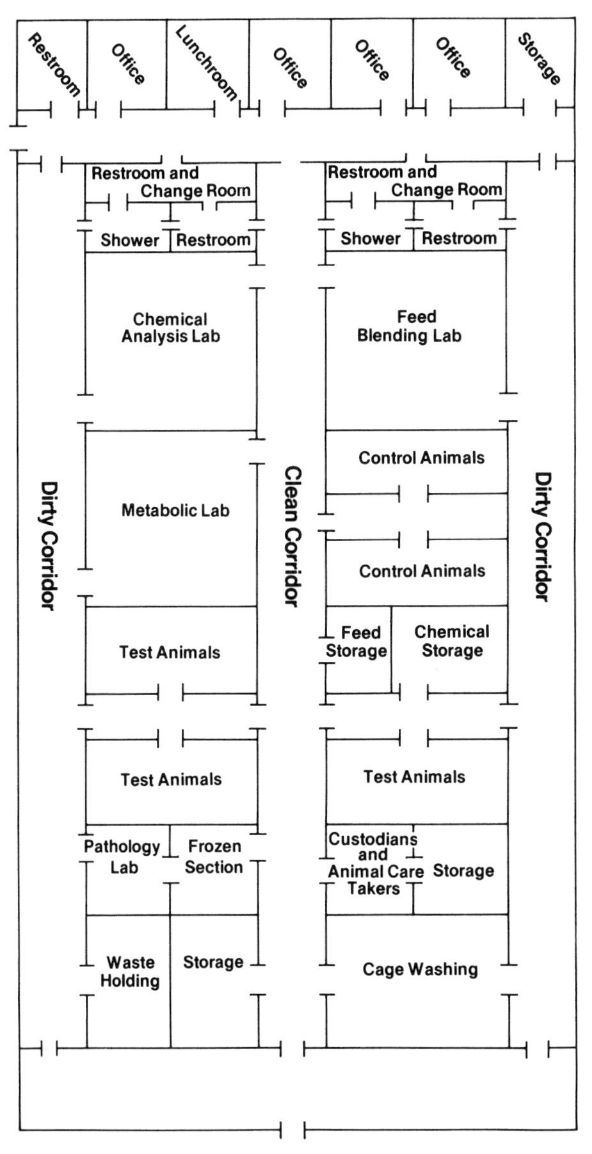

FIGURE 6.1. Schematic of the cancer research laboratory.

air is distributed to all rooms through ceiling diffusers and air is exhausted through laboratory hoods in the research area and through exhaust grills in the nonrestricted areas.

A "dirty" corridor on the perimeter of the restricted area is used for the movement of contaminated cages, animals, and feed. A clean corridor extends down the middle of the research area and connects to the dirty corridor in the rear of the building. Generally, the flow of materials and equipment is from clean to dirty corridors. All laboratories have doors that open to both the dirty and clean corridors but that remain closed during most operations. Other rooms in the restricted area are devoted to cage washing, control animal holding, chemical and animal feed storage, animal waste holding, and animal testing and holding.

Feeding studies involving rats and mice are conducted in the animal testing and holding rooms. These rooms are specially designed to prevent cross-contamination between holding areas containing animals of different experimental groups. Dosing or feeding of all animals takes place in this area.

The laboratory director is very conscientious about the health and safety of his staff and takes pride in the fact that there has never been a known work-related illness or injury among his personnel. Recently, the laboratory director asked the campus health and safety staff to conduct a comprehensive health and safety review of the laboratory, with initial emphasis on the chemical analysis and metabolic laboratories.

Nature of the Problem

CHEMICAL ANALYSIS LABORATORY

A walk-through survey of the chemical analysis laboratory was conducted by the campus health and safety staff. This laboratory measures 20 × 20 × 8 ft. Upon receipt of the concentrated BBD, the technician dons protective gloves and a laboratory coat, places the new container of dye in the laboratory hood, checks the labels for appropriate contents, and opens the shipping container. A physical inspection is made of the inner container for damage and/or leakage. Defective or leaking containers are immediately reported to the supplier who arranges for their return. If no problems are noted, a small amount of the material is removed from the container and weighed on a balance inside the laboratory hood. A sample of the dye is then assayed for chemical purity. After purity is assured, the technician prepares aqueous solutions of 0.5, 1.0, 2.5, and 5.0% (by weight). The solutions are reanalyzed to assure accuracy of the dilutions. The dilutions are sealed in double containers and sent to the feed blending laboratory. After the blending laboratory staff prepare batches of dosed feed, approximately 150 g of each batch is returned to the chemical analysis laboratory for assay.

The chemical analysis laboratory is equipped with an eyewash and deluge shower. The laboratory has a 10-lb ABC fire extinguisher located near the clean corridor door. The laboratory hood is an auxiliary air type that was originally selected for energy conservation and personnel safety. The auxiliary air laboratory hood has an integral air-grill canopy above the sash opening through which outside air is supplied by a separate fan, which is located on the roof. The outside air flows down the front of the hood face and is drawn through the work opening by the hood exhaust fan, which is also located on the roof. The purpose of auxiliary air hoods is to use some unconditioned or partly conditioned outside air for exhaust air, instead of exhausting 100% conditioned building air. The objectives of such a system are to reduce air-tempering costs, but another and probably most important feature of the auxiliary air hood is that it continuously bathes the technician with clean fresh air, hopefully minimizing the possibility of contaminants escaping from the hood into the technician's breathing zone. These objectives may or may not be met.

The technicians using this laboratory hood stated that during extremes in outdoor temperature, the incoming air is either too cold or too hot. To correct this problem, the technicians taped a piece of cardboard over the auxiliary inlet grill to close off this air supply. As a result, the hood draws only conditioned air from the building, thereby eliminating the discomfort problem.

The technicians also complained that during the weighing of the dry dye, the air velocity through the hood was so great that sometimes a portion of the material is blown off the balance. To correct this problem, the technicians usually turn the hood off during the weighing procedure. All BBD weighing is done with the hood sash at 6 in., although there is occasionally a need for the hood sash to be wide open.

Inspection of the roof area where the laboratory hood exhaust and supply fans are located revealed that the fans are approximately 3 ft apart. Both stacks terminated approximately 1 ft below the roof parapet.

The laboratory was neat and orderly, except for an accumulation of dust around the laboratory side of the door jam leading to the clean corridor. The technicians appeared to be well trained. All attend regular safety training sessions, and safety meetings are held weekly. The technicians have excellent safety records and stated that they had been informed by the laboratory director of the hazards of working with BBDs.

METABOLIC LABORATORY

Work in the metabolic laboratory involves studies to determine the metabolic pathways and accumulation rates of the chemical test agents. In addition to standard biochemical methods for determining metabolic pathways, radioactive-labeled materials added to dosed feed stock are also used. The BBD is added to the feed in the blending laboratory, and a portion of the dosed feed stock is transferred to the metabolic laboratory, where some radioactive BBD is added and used for certain metabolic studies. The label currently under investigation is carbon-14 (^{14}C). Measurements of ^{14}C-labeled compounds in blood and waste products are accomplished using scintillation techniques. Collections of animal blood and wastes are made by technicians, and the samples are returned to the metabolic laboratory where the radioactive content of the wastes is determined. Approximately 100 mC. of ^{14}C are used each month.

Radioactive materials work is performed in a specific area of the room. The laboratory door from the clean corridor is posted with a standard "Caution—Radioactive Materials" sign. Bench tops, where the radioactive materials work is performed, are covered with absorbent paper; they are also marked with caution signs. The stock radioactive materials are stored behind a lead shield in a laboratory hood located at one end of the room. The laboratory director has had formal training in radiation safety procedures, and the technicians have been instructed by him. When the radioactive tracer studies were initiated, the laboratory director met with the technicians and explained that the radioactive materials to be used were low-energy isotopes; if they followed the same safe work procedures as used when handling chemical carcinogens, including the use of protective wear where necessary, there would be no additional hazards. The staff was also informed that they must keep accurate records of the amounts of radioactive material administered to each animal.

Questions

1. What measurements, if any, should be made of the analysis laboratory hood performance, and what type of equipment would you use? Are there any indications of poor ventilation conditions in this laboratory?
2. What potential problems might be caused by the locations of the supply and exhaust fans on the roof, and what additional studies would you recommend?

3. Is an in-depth study of the metabolic laboratory needed? If so, how will you conduct it? What instruments are needed? What will be evaluated?

6.2. EVALUATION RESULTS

The fact that the technicians must block off the auxiliary air grill to achieve temperature comfort indicates that the incoming air is probably not tempered. This particular laboratory hood is designed to draw approximately 60% of make-up air from the outside. In addition, the laboratory ventilation is designed around the hood. Consequently, blocking off the auxiliary air grill causes the hood to draw all make-up air from the laboratory. This situation may cause the laboratory to be under negative air pressure; air could be drawn from the dirty corridor through the laboratory and into the hood. Normally, the air pressure in the laboratory should be greater than that in the dirty corridor. Typically, air flow in a laboratory facility of this type should always be from clean to dirty. Therefore, the clean corridor should be overpressured and the dirty corridor underpressured in comparison to the laboratory.

Based on the observations made during the walk-through survey, the industrial hygienist conducted air-flow measurements of the laboratory hood. The measurements were made using a Model 415B-3 Kurz heated-wire anemometer, which was calibrated using a Model 1164 Kurz wind tunnel. The calibration was performed according to the manufacturer's instructions, and the calibration procedure was traceable to the National Bureau of Standards. The overall accuracy of the anemometer was ±5% below 250 fpm and ±2% in the range of 250 to 2000 fpm.

Measurements of the hood's air flow were obtained according to procedures described in the ACGIH *Industrial Ventilation Manual*. Upon inquiry, it was determined that some hood work required a full open-sash position. After the hood face fully open dimensions were determined, air velocity measurements were made at the center of each 1-ft^2 opening. A total of 12 measurements were made. The maximum velocity was measured at each point by rotating the sensing probe to assure that the air crossed the sensing probe at a right angle. Additionally, there were no air-flow obstructions inside the hood that could cause nonuniform air-flow patterns. All measurements were made with the auxiliary air blocked off and full open sash. The air-velocity measurements obtained are shown in Figure 6.2.

Before the weighing procedure began, the laboratory hood sash was lowered, leaving approximately a 6-in. work opening. The weighing procedure was then observed in detail while the hood was on. A definite problem was observed. Not only was the dye blown off the balance, but it was also difficult to transfer the dye material from the shipping container to the balance. In the transferring process, some of the dye was blown off the spatula.

Subsequently, additional air-flow studies of the laboratory hood at the 6-in. work opening height were conducted. Air-flow measurements were taken every 12 in. across the face opening. A total of four measurements were taken, and the results are also presented in Figure 6.2.

Smoke tube tests were conducted with the hood turned off and on. In the off condition, smoke was released in the laboratory near the dirty corridor door and near the clean corridor door. In both cases, the smoke traveled out of the laboratory into the nearest corridor. With the hood on (the auxiliary grill still blocked) and smoke released at the same locations as before, smoke always traveled toward the hood and away from either corridor.

Air-flow measurements were made of the supply air in the laboratory. The calibrated Kurtz Model 415B-3 heated wire anemometer was used to measure the air velocity from all four ceiling air outlets. Each ceiling outlet measured 2 ft by 2 ft. A

48 in.

143	149	153	141
152	160	162	150
149	150	148	143

30 in.

728	780	771	721

6 in.

FIGURE 6.2. Velometer readings (fpm) at the hood face: (a) sash in the full open position, (b) sash open to 6 in. height.

total of four measurements were made on each outlet. The average results were 32, 35, 35, and 38 fpm for the four outlets.

Since the auxiliary air fan and the laboratory hood exhaust fan are located within 3 ft of each other on the roof, the potential exists for exhaust air to be reentrained into the supply air system. Adding to this potential problem is the fact that the exhaust duct terminates below the roof parapet. Depending on climatic conditions, this situation could cause the pooling of exhaust air below the parapet, and the exhaust air could be drawn into the auxiliary air system. Because of this situation, a tracer study of the exhaust and auxiliary air supply system was conducted. The climatic conditions at the time of the study were: dry bulb temperature, 21°C; relative humidity, 60%; and wind, 10 mph from the south.

Pure liquified sulfur hexafluoride (SF_6) was obtained in a cylinder containing 8.2 kg (1370 L at NTP). To quantify the amount of the tracer gas released, the SF_6 cylinder was connected to a rotameter (Dwyer No. UBA-7B Visa Float) calibrated from 2.4 to 23.6 Lpm. A stopwatch was used to time each release of tracer gas.

The analytical instrument used to monitor SF_6 was a MIRAN-1A portable gas analyzer (Foxboro-Wilks) equipped with an integrated sampling pump, 3-m sampling hose, and particulate filter. A strip-chart recorder (Linear Instruments) was attached to the MIRAN-1A to record variations in SF_6 concentrations with time. The curve produced on the strip chart was traced with a planimeter in order to calculate the total area under the curve.

The MIRAN-1A was calibrated using a Foxboro-Wilks closed-loop calibration pump. The manufacturer's operating procedures were followed. Optimum operating conditions were: wavelength, 10.7 μm; slit width, 1 mm; path length, 20.25 m; and response time, 1 sec. These conditions gave a lowest detectable level of 25 ppb.

The SF_6 was released at a constant predetermined rate for a predetermined time, inside the laboratory hood. The sampling point was located directly below the auxiliary air supply grill and was moved about the middle portion of the grill in order to obtain an approximate average reading. The MIRAN-1A was operated until the baseline was regained.

An increase at any point on the strip chart *Y* axis of 1.0 cm equaled a 10% increase in scale (S) or output. An increase along the *X* axis of 1.0 cm equaled an increase of 1/2 min.

From calibration data a regression analysis was performed, and the slope of the calibration curve was determined to be 0.073 μL SF_6 per 1% S. This volume of SF_6 was converted to a concentration, knowing that the MIRAN-1A with calibration pump has a volume of 5.64 L. In this case, 0.073 μL SF_6 in the MIRAN-1A system corresponds to 0.013 ppm SF_6. Therefore, each 1% increase in scale is equivalent to an increase of 0.013 ppm SF_6.

The total volume of air from which the MIRAN-1A drew its sample was determined by using the sampling time and the known hood flow rate of 1500 cfm (42.5 m^3/min). With this information, the total SF_6 reentrained was then calculated. It should be assumed that SF_6 is well mixed throughout the system, that is, the sample seen by the MIRAN-1A is representative of the average air concentration. However, this seems reasonable due to the mixing time and the movement of the sampling probe at the air inlet duct. The percentage of SF_6 reentrainment was then determined. Data from three trials are presented in Table 6.1.

In the metabolic laboratory a survey was conducted for radiation contamination using a Warrington Model 2590 Cutie Pie. No contamination was found. In addition, wipe samples were collected from selected locations in the laboratory, since detection of low-energy beta emission using survey instruments is difficult. The decision to collect wipe samples was also based on the fact that contamination surveys had not been previously conducted in the laboratory.

The samples were taken using No. 4 Whatman filter disks, which were wiped over a 100 cm^2 surface area as per Nuclear Regulatory Commission regulations. The filter paper was then placed in a scintillation vial, and 5 mL of scintillation fluid was added. The samples were counted for 1 min each, in a Tracor Analytic Mark IV scintillation counter. Prior to counting the samples, a calibration standard containing 0.01 μCi of ^{14}C was counted for 1 min in the scintillation counter and resulted in 1.78×10^4 counts. Figure 6.3 shows a diagram of sample loca-

TABLE 6.1
Tests for Exhaust Air Reentrainment

Trial	Area under Curve	Sample Time	Volume of SF_6 Released
1	10 cm^2	3 min	0.5 L
2	18 cm^2	2 min	0.5 L
3	40 cm^2	3 min	1.0 L

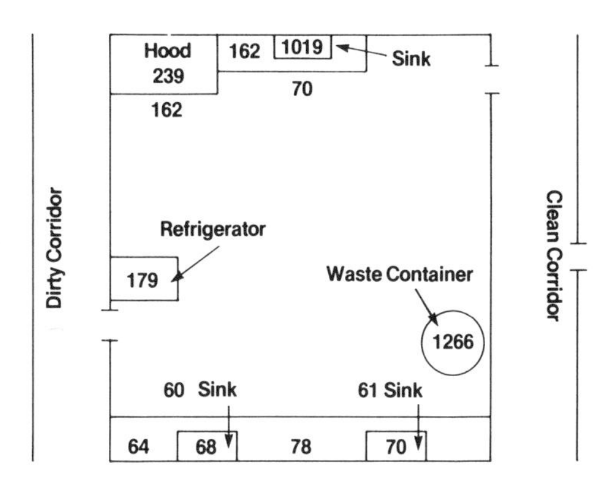

FIGURE 6.3. Diagram of results of wipe samples for radioactivity (gross cpm) taken at locations where results are shown.

tions and the results of the counted samples. A blank filter was counted and a background reading of 64 counts/min was obtained.

Although low energy beta emitters have a relatively short travel distance in air and usually do not present external exposure problems, the storage of ^{14}C behind the lead shield was also evaluated. The concern was for the possibility of bremsstrahlung (X radiation) production when a beta emitter is stored in a high-Z-number shield.

The following equation was used to determine if there was a potential hazard associated with bremsstrahlung radiation.

$$f = 7 \times 10^{-4} ZE$$

where f is the fractional bremsstrahlung production, 7×10^{-4} the bremsstrahlung proportionality constant, Z the atomic number of shielding, and E the maximum beta energy for ^{14}C in MeV. In the above equation, f represents the fraction of emitted beta energy that appears as bremsstrahlung radiation, in this case, 8.9×10^{-3} or 0.89%. Values for E and Z can be found in a handbook such as *The Radiological Health Handbook*. The "effective gamma ray" activity was then calculated using the equation

$$A_g = fA_b$$

where A_g is the effective gamma-ray activity in curies, f the fractional bremsstrahlung production, and A_b the beta activity in curies of ^{14}C. Since the A_b value for ^{14}C is 0.1 Ci, the A_g produced by that beta activity and the properties of the lead shield is 8.9×10^{-4} Ci. To calculate the exposure rate, the following equation was used.

$$R_1 = 0.5\ CE$$

where 0.5 is a constant, C the activity in curies, E the energy in MeV, and R_1 = roentgen/hr at 1 meter. In this case, the predicted exposure at 1 m due to bremsstrahlung radiation would be 0.07 mR/hr or mrem/hr. (Since the radiation in question is X radiation, 1 mR = 1 mrem.)

Questions

1. Based on the ventilation measurements and observations, what conclusions and what recommendations (if any) would you make regarding ventilation?
2. What do the data from the SF_6 tracer study indicate? What recommendations would you make based on these results?
3. Based on the wipe-sample results, what are your conclusions and recommendations?
4. Does storage of the ^{14}C in the lead shield present an exposure problem?

6.3. CONTROL STRATEGIES

Calculations of the laboratory hood air volume from data in Figure 6.2 revealed that 90,000 cubic feet per hour (cfh) (STP) of air is exhausted through the hood when the sash is fully open. Calculation of the laboratory supply air volume from measurements of air velocity at the supply grills indicated that there are 33,600 cfh provided to the laboratory. This indicates that the laboratory is under significant negative pressure. The situation is primarily caused by the technicians practice of blocking off the auxiliary air grill. As described earlier, the auxiliary air system is designed to provide 60% (54,000 cfh) of the make-up air needed for the hood exhaust. If the auxiliary air system was used as designed, only 40% of the air needed for hood exhaust (36,000 cfh) would be required from the laboratory. This amount is slightly more than the quanitity provided by the laboratory supply system (33,600 cfh). This deficit of supply air (2400 cfh) creates a slight negative air pressure in the laboratory, which is the intended design, and promotes air flow from the clean corridor into the laboratory.

These calculations were confirmed in the smoke-tube tests. Smoke released at the dirty corridor door, when the grill was blocked and hood on, was drawn from the dirty corridor through the laboratory toward the laboratory hood. After these calculations were made, smoke was released at the dirty corridor door with the grill unblocked, and the smoke was drawn through the laboratory door and down the dirty corridor. This demonstration proved that the laboratory hood should be operated as designed, with the auxiliary air system unblocked. To solve the technicians' discomfort problem, the auxiliary air should be temperature adjusted during extreme cold and hot weather.

The average air velocity with a 6-in. sash opening was 750 fpm (Figure 6.1). This is a very high face velocity, especially when weighing of dry material is to be performed. The solution to this problem is to reduce the face velocity of the hood so that dry weighing can be achieved without turning the hood off. It also may be appropriate to recommend that an air shield or glove box be provided for transferring and weighing dry materials. This air shield or glove box would provide a "quiet zone" for weighing, while the air movement around the shield or box would exhaust escaping or spilled material. Smoke-tube tests should be conducted to determine air-flow patterns around these devices before they are permanently installed.

Usually, dry weighing can be performed satisfactorily at velocities of 200 to 250 fpm. NIOSH and former ACGIH (1980) recommendations were that chemical carcinogen laboratory hoods have an average minimum face velocity of 150 fpm. (In 1982 ACGIH recommendations were lowered and are discussed in the *Other Solutions* section.) Therefore, on first observation it seems quite simple to lower the hood face velocity by slowing down the fan. However, it should be noted that the American Society for Heating, Refrigeration, and Air Conditioning Engineers (ASHRAE) recommends a minimum of ten air changes per hour in laboratories. Since 33,600 cfh of fresh air is supplied and the room volume is 3200 ft^3, 10.5 air changes per hour take place when the system is operating as designed. Consequently, a hood velocity reduction would also require that supply air be reduced in order to balance the system, an unwanted result.

From Figure 6.1, the average face velocity is calculated as 150 fpm, in the full open-sash condition. This face velocity meets the NIOSH recommendations; although the laboratory air change rate is slightly more than needed (10.5), the hood velocity should not be lowered because the hood is sometimes used in a full open-sash position, and any lower face velocity would not meet NIOSH recommendations of 150 fpm. While the face velocity of 150 fpm is certainly satisfactory, the 200 – 250 fpm condition for successful dry weighing can be achieved by lowering the sash to a point that provides maximum face velocity without weighing interference. A sash height of 20 in. would provide a face velocity of approximately 225 fpm. After a simple test to determine if the face velocities at this sash height interfere with weighing, the laboratory hood should be marked to indicate a 20-in. height opening, and the technicians should be instructed that the 20-in. height is to be used during weighing.

The smoke-tube tests indicate that the laboratory is under positive pressure when the hood is turned off. Because the hood is the only means of exhaust from the laboratory, it should be obvious that the laboratory would be under positive pressure when the hood is off and the fresh air supply system is on. The first indication that the laboratory is sometimes under positive pressure is the accumulation of dust on laboratory side door jamb of the clean corridor door. If the laboratory were under constant negative pressure at that point, dust accumulation, if any, would be expected on the corridor side of the door jamb. In this laboratory, the question of whether the observed dust is dye material or normal dust is pertinent. Because of the conditions found and the technicians' practice of turning off the hood, it should be recommended that dust samples be tested for dye material.

The data provided in Table 6.1 can be used to calculate the percentage of SF_6 that is reentrained in the hood exhaust air. An example calculation for trial one is as follows. First, the concentration of SF_6 in the fresh air (reentrained) (C_R) is calculated.

$$C_R = 10\ \text{cm}^2 \times \frac{5\%\ \text{S min}}{1\ \text{cm}^2} \times \frac{0.13\ \text{ppm}}{1\%\ \text{scale}} \times \frac{1}{3\ \text{min}}$$

$$= 0.22\ \text{ppm SF}_6$$

Next, the volume (V_R) represented by 0.22 ppm SF_6 is calculated, given that the intake duct flow rate is 25.5 m^3/min and sample time is 3 min.

$$V_R = \frac{25.5 \text{ m}^3 \text{ air}}{\text{min}} \times 3 \text{ min} \times \frac{0.22 \text{ L SF}_6}{\text{m}^3 \text{ air}} = 0.02\text{L SF}_6$$

Then, the percentage of the released SF_6 that was reentrained ($\%R$) is calculated.

$$\%R = \frac{0.02 \text{ L SF}_6 \text{ captured}}{0.5 \text{ L SF}_6 \text{ released}} \times 100\% = 4\%$$

It may be appropriate to conduct additional tracer studies to determine a "worst-case" situation with respect to environmental conditions. However, the data indicated reentrainment on all three trials of 4, 6, and 6%, respectively. Although tracer studies conducted under different climatic conditions might indicate less or more reentrainment, additional studies are not necessary to justify corrective action. Since there are no established safe levels of exposure to BBD, the data obtained from the tracer study are sufficient to recommend changes in the exhaust and auxiliary air locations.

One recommendation which could be made is to redesign and equip the exhaust system with a high-efficiency particulate air filter (HEPA). This should be done as a matter of good practice to prevent toxic chemicals, especially carcinogens, from contaminating community air and also to prevent reentrainment into other air intake systems. At the very least, the exhaust or supply for this laboratory hood should be relocated to obtain as great a vertical and horizontal separation as possible. In addition, the exhaust outlet should be extended above the roof parapet and provided with a high-velocity nozzle discharge so that the exhausted air is directed upward. This should assure maximum dilution of the exhaust air.

A word of caution is in order when modifying existing air systems. Additional measurements of the laboratory hood face velocities and a check of the mechanical capabilities of the existing exhaust fan must be made before and after installing filters or other restrictions to the system. Often this will require a complete redesign of the systems. The capacity of the fan and motor may not be capable of the additional static pressure created by adding filters to the system. Once the changes in the ventilation system have been made, additional tracer studies should be conducted to verify correction.

The results for wipe samples from the metabolic laboratory shown in Figure 6.3 require adjustment. First, as was stated earlier, the scintillation counter yielded a count of 1.78×10^{-4} counts/minute (cpm) for a source of 0.01 μCi of ^{14}C. Since 0.01 μCi should equal 2.22×10^{-4} cpm, a correction factor of 1.25 was applied to each reading from the lab. In addition, the background count of 64 dpm was subtracted from the gross count prior to multiplying by 1.25. The resulting net dpm data are shown in Figure 6.4.

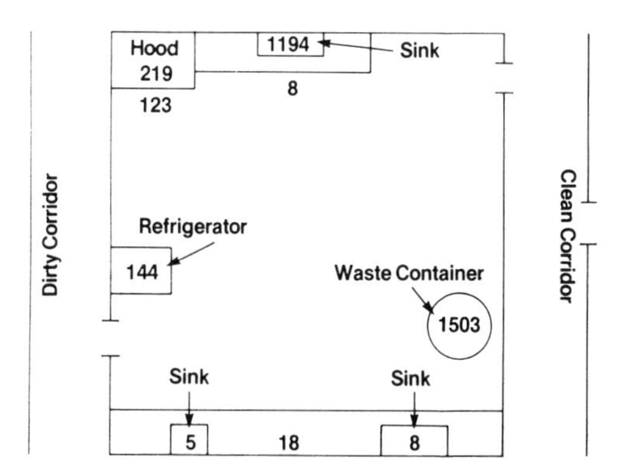

FIGURE 6.4. Diagram of results of wipe samples for radioactivity (net counts/min) taken at locations where results are shown.

The net results were compared to the recommendations of the National Committee on Radiation Protection (NCRP). This report recommends that contamination levels be less than 100 dpm (above background) per 100 cm^2 of surface area. Any levels above 100 dpm/100 cm^2 is considered to be significant contamination, and immediate decontamination is warranted. Using the NCRP recommendations as a guide, five sample locations had significant contamination and should be decontaminated immediately. Although the other sample results do not indicate that immediate decontamination is necessary, the fact that some contamination was found in four other samples is indicative of improper handling techniques. Furthermore, it is obvious that radiologically contaminated materials are being thrown away and flushed down the sink. Consequently, the laboratory employees should be given training that emphasizes their responsibility, the danger to their health, and possible disciplinary actions that could result from further improper disposal techniques.

The results obtained from the bremsstrahlung calculations were compared to the external occupational radiation exposure limit standards of the Nuclear Regulatory Commission and the Occupational Safety and Health Administration. Assuming a worse case situation, in which technicians would be exposed 8 hr/day at 1 m, the whole body exposure would be 0.56 mrem per day or 36 mrem per calendar quarter. Based on these calculations, the bremsstrahlung production does not yield a presumed whole-body exposure to technicians that is excessive (OSHA standard is 1.25 rem/quarter). Even though the bremsstrahlung radiation level is low, the ALARA concept (as low as reasonably achievable) requires that ionizing radiation exposures be reduced whenever practical. In this case, it was recommended that the lead shielding be replaced with a material having a low *Z* number, such as 1/4 in. acrylic plastic.

External personnel monitoring using film badges is not necessary nor required. The Nuclear Regulatory Commission standard requires external monitoring if 25% (0.3 rem) of the whole-body exposure limit can be received in a calendar quarter.

Based on the results of the samples taken and the information obtained on the walk-through survey, the following should be done for the metabolic laboratory:

1. Decontaminate the laboratory.
2. Resurvey the laboratory to assure decontamination.
3. Conduct a radioisotope-handling-techniques course for all metabolic laboratory workers.
4. Establish a daily laboratory contamination survey and record-keeping program.
5. Because significant contamination was found in the laboratory, it is appropriate that urine samples of all metabolic laboratory personnel be checked for ^{14}C contamination. The sampling, analysis, and interpretation of the results should be based on recommendations of the International Council on Radiation Protection (ICRP).

Alternative Solutions

For the exhaust hood in the chemical analysis laboratory, adjustment of interior exhaust slots may also solve the face velocity problem. These slots are located at top and bottom of the back panel of the hood; they control the location of the flow of air through the hood. By adjusting the bottom slot to a smaller opening, the velocity in the lower half of the hood can be decreased. However, the top slot must be opened wider to compensate for static pressure losses. After slot adjustments are made, the face velocity should be rechecked.

In weighing of dry materials, slot adjustments may or may not be helpful, and a trial and error approach is necessary. Normally, processes that take place near the bottom or top of the hood will be most affected by respective slot adjustments. For weighing, most of the activity probably takes place near the middle of the hood, and would be least affected by changes.

Recently, the ACGIH (1982) revised their recommendations for appropriate laboratory hood face velocities. These revisions are based on an American Society of Heating, Refrigerating, and Air Conditioning Engineers (ASHRAE) report (Caplan and Knutson), which indicated that hood face velocities in the range of 60–150 fpm (or 60–150 cfm/ft^2 of opening) were sufficient to provide good control of contaminants. The upper end of this range is used when movement in front of the hood is greatest. The report stated increasing the face velocity increases the eddying effect, thereby causing contaminants to be brought to the breathing zone of the worker.

A redesign of the hood in this case study to meet the above recommendation is certainly feasible and warranted if movement at the hood is minimal. In this case, the hood fan speed would have to be reduced. Although make-up air would not have to be reduced, it would certainly reduce the discomfort of the technicians and perhaps negate the need to temper the make-up air. At any rate, careful design would be warranted, and after changes are made, a tracer study should be performed. This is to assure that concentrations of tracer at breathing zone meet a preset criteria when the tracer is released in the hood. Caplan and Knutson recommend a breathing zone concentration of < 0.1 ppm when SF_6 is the tracer.

Provided that changes can be made to the current system, perhaps the best solution would be to reduce flow to about 100 fpm at some reduced sash height that would still allow weighing to be done. This may reduce total flow such that auxiliary air would not require tempering. Problems with auxiliary air hoods are now well known and discussed in ACGIH (1982); however, many auxiliary air hoods still exist in laboratories throughout the nation, and industrial hygienists will have to deal with the problems that they present.

Urinalysis for the presence of ^{14}C was recommended for the metabolic laboratory workers. Biological monitoring might also be recommended for the chemical analysis workers due to their potential exposures to benzidine or other metabolites. Unfortunately, the results of such surveys may be difficult to interpret. For ^{14}C, the measured values are compared with ICRP standards (1979). However, for benzidine the best comparison is the employee's own baseline. In this case, a preemployment baseline, or one taken while all control systems are properly functioning, would be appropriate. Under these conditions, metabolites would probably be undetectable. Consequently, any detection of metabolites at a later date would indicate failure of controls, possibly including failure to observe safe work practices. Positive metabolite results should, at the very least, encourage the workers to observe all health and safety precautions.

Although not the subject of this case study, other portions of the laboratory safety program are important to consider. These include use of protective gloves, emergency operating procedures, routine physicals, documentation of training, work practices, and the like. Information for each of these is available in National Toxicology Program guidelines and in Walters.

REFERENCES

American Conference of Governmental Industrial Hygienists. *Industrial Ventilaton Manual*, (16th and 17th eds.), ACGIH, Cincinnati, OH, 1980, 1982.

American Conference of Governmental Industrial Hygienists. *Workplace Control of Carcinogens*, ACGIH, Cincinnati, OH, 1977.

Bureau of Radiation Health. *Radiological Health Handbook*, U.S. Department of Health, Education, and Welfare, Rockville, MD, 1970.

Caplan, K.J., and G.W. Knutson, Laboratory fume hoods: a performance test, *ASHRAE Transactions*, 84: Part 1, 1978.

Gollnick, D.A. *Basic Radiation Protection Technology*, Pacific Radiation Press, Temple City, CA, 1983.

International Commission on Radiation Protection. *Annals of the ICRP*, Publication No. 25, Pergamon, Oxford, 1979.

National Council on Radiation Protection and Measurements. *Safe Handling of Radioactive Materials*, Report No. 30. U.S. Department of Commerce, Washington, DC, 1964.

National Institute for Occupational Safety and Health. *Health Hazard Alert: Benzidine-,* o-*Tolidine-, and* o-*Dianisidine-Based Dyes*, U.S. Department of Health, Education, and Welfare, Cincinnati, OH, 1980.

National Institute for Occupational Safety and Health. *Special Occupational Hazard Review for Benzidine-Based Dyes*, U.S. Department of Health, Education, and Welfare, Cincinnati, OH, 1980.

National Toxicology Program. *General Statement of Work for the Conduct of Acute, Fourteen-Day Repeated Dose, 90-Day Subebrionics, and 2-Year Chronic Studies in Laboratory Animals*, Department of Health and Human Services, Research Triangle Park, NC, July, 1984.

Nuclear Regulatory Commission. Rules and Regulations, *Energy*, Title 10, Chapter 1, Code of Federal Regulations, Washington, DC, 1983.

Department of Occupational Health and Safety, University of Alabama at Birmingham. *Radiation Safety Procedures Manual*, 1983.

Walters, D.B. (Ed.). *Safe Handling of Chemical Carcinogens, Mutagens, Teratogens, and Highly Toxic Substances*, Vol. 1 and 2. Ann Arbor Science-Butterworth Group, Ann Arbor, MI, 1980.

7

ALUMINUM REDUCTION POTROOM OPERATIONS

MICHAEL C. RIDGE

7.1. PROBLEM RECOGNITION

Process Description

Aluminum is produced by The Hall–Heroult process, an electrolytic reduction of aluminum oxide (alumina), which was developed in 1886. A fused salt bath, usually just referred to as "the bath," is used to dissolve the alumina and allow the passage of an electric current. This bath is cryolite, Na_3AlF_6, which is able to dissolve nearly 10% alumina by weight.

The reduction cell cathode consists of a large, shallow, steel box lined with a layer of carbon and containing the bath and the molten aluminum. The steel box is termed the cathode, although the aluminum metal, which is heavier than the bath and sinks to the bottom of the cell, is the true cathode. Often layers of refractory brick are placed between the carbon lining and the steel shell to insulate it from the 950°C bath.

The reduction cell anode consists of one or more large monolithic blocks of carbon suspended above the cathode and immersed in the bath. As the alumina is reduced to aluminum, the liberated oxygen migrates to the surface of the anode, where it reacts to form carbon dioxide and lesser amounts of carbon monoxide. This reaction continuously consumes the anode surface. The overall reaction is:

$$Al_2O_3 + 1.5C \rightarrow 1.5CO_2 + 2Al$$

Both the cathode and anode carbon are produced by grinding coal and coke and mixing them with coal tar pitch, or sometimes petroleum pitch, as a binder. The resulting paste is shaped and baked to produce the anode and cathode materials.

In forming the cathode, the lining is shaped in the cathode shell and slowly "baked-in" when the cell is first placed in service. The process is carefully controlled to produce a strong lining, which should have a service life of 3 or more yr. Most of the volatile materials in the coal and pitch are driven off or carbonized during this bake-in period.

In the case of the anode two processes are used: the prebake process and the Soderberg process. In the prebake process, the anode is fabricated in an anode plant, where a paste of petroleum coke and pitch is mixed, shaped, and baked. The baked anode is then fitted with a conductor rod, which is used to suspend it above the cathode from the anode bus bar. Prebake cells typically use six or more prebaked anodes. Each anode is replaced as it is consumed by reaction with oxygen.

The Soderberg anode is baked in place above the cell, using the heat generated by the cell. The anode paste is continually added to a form or shell suspended above the cathode. As it bakes, it forms a solid block of carbon. The anode paste is added to the top of the shell at a rate equal to the consumption of the bottom surface by the reduction reaction. The Soderberg anode and its shell are suspended from the anode bus by conducting rods called studs or pins, which are inserted into the paste and bake into the carbon block. As the Soderberg anode is consumed, the studs nearest the bath are removed and replaced with new studs set in the uppermost paste layer. Therefore, there are studs at different levels in the baking anode. In one design, the studs are set horizontally into the anode. This is called the horizontal stud Soderberg, or HSS, cell. In the other design, the studs are set into the cell vertically, and the cell is called the vertical stud Soderberg, or VSS, cell.

In the anode baking process, the volatile materials in the coke and pitch are driven off or carbonized, just as they are in the cathode bake-in period. However, small quantities of complex hydrocarbons remain trapped or bound in the anode structure. These are released as the anode is consumed. Both the volatile materials, which are driven off during baking, and these trapped materials, which are released when the anode is consumed, contain significant amounts of polynuclear aromatic hydrocarbons, or PNAs. These are so termed because they are made up of linked six-carbon aromatic rings. For prebake operations, most of the PNAs are driven off in the anode plant. However, in a Soderberg plant, significant quantities of polynuclear aromatic hydrocarbons will evolve from the anode tops of the operating reduction cells.

Each of the three types of reduction cells is equipped with distinctive ventilation controls to collect various contaminants, which are released during normal cell operation. However, none are totally effective, and most of the ventilation designs do not operate at their maximum capture efficency during certain phases of normal cell operations. The vertical stud Soderberg ventilation is least affected by operating cycle changes, but the top of the anode is typically not hooded at all, so that

PNAs are released directly into the workroom air from the anode top. Figure 7.1 shows each cell type with its ventilation collection system.

The plant that is the subject of this case study is a VSS plant of about 100,000 tons/yr production. The following information is specific to this plant and is typical of the background data required in planning an exposure evaluation.

In this plant there are 300 cells arranged in two potlines. (See Figure 7.2.) A potline consists of a group of cells in direct current electrical series. The cell voltage is approximately 4.2 V, which means that the voltage between one end of the potline and the other is $150 \times 4.2 = 630$ V. The line operates at 106,500 A. The cathode for each cell is about 29 ft long by 15 ft wide. The anode is about 23 ft long by 7½ ft wide, and is equipped with a skirt-type fume collection system (see Figure 7.1). The potline buildings are provided general ventilation by large fans that accomplish 30 air changes per hour. This cools cathodes and provides some air movement at the work level of

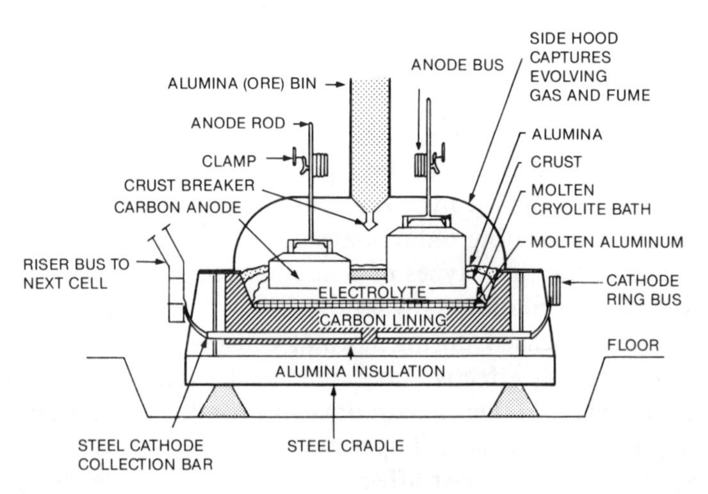

FIGURE 7.1a. Cross section of prebake aluminum reduction cell.

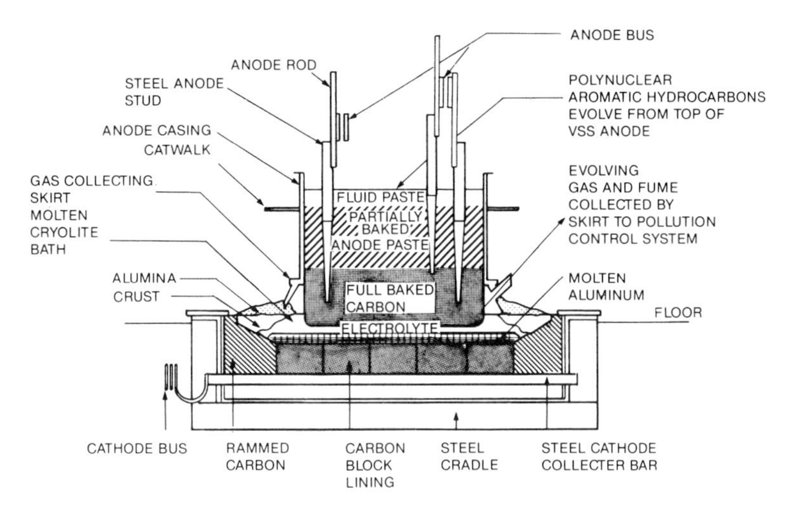

FIGURE 7.1*b*. Cross section of VSS aluminum reduction cell.

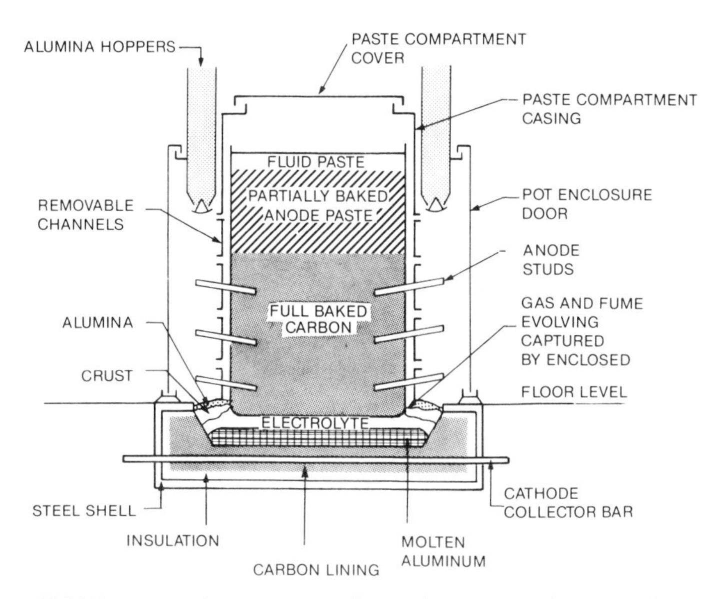

FIGURE 7.1c. Cross section of HSS aluminum reduction cell.

FIGURE 7.2. Cut-away of potroom of VSS plant in this case study.

the potrooms. The fresh air absorbs heat as it passes over the cathode shells, so the potrooms are quite hot, especially in summer.

There are four shifts that operate the potlines on a rotation that allows 24 hr a day coverage with one shift always off. On each shift there are 20 cell operators, 2 for each group of 30 cells. There are two 3-person, stud pulling crews per shift. One briquetter, one fluoride truck driver, and a 3-person tapping crew complete the potline crew for each shift.

Shortly after beginning the shift, the crust on each cell is broken open using a small vehicle called a crust breaker. This mixes the alumina charge from the last shift into the bath. The operator will normally "rake" the bottom of the anode while the cell is broken open. The anode-cathode distance is then adjusted by lowering the anode to compensate for the amount of consumed carbon, restoring the cell voltage to 4.2 V. The other cell operator uses an ore truck to fill the cells by placing a charge of alumina onto the refrozen crust. After all cells have been "ored," the cell operators must sweep the ore against the skirt (see Figure 7.1) to seal the cell and insure maximum collection of contaminants released by the cell. This is followed by housekeeping and maintenance tasks. Once these are completed, the cell operators can take a prolonged break in their work area until the next cycle of crust breaking and ore replenishment, which occurs halfway through the shift.

Occasionally a cell will exhaust its alumina charge before the next scheduled cycle of crust breaking; when this occurs the cell operator must perform an unscheduled crust break in order to recharge the cell with alumina and restore the cell to proper operation. A klaxon horn and indicator light alert the cell operators if the cell exhausts its alumina charge; hence this event is called a "light" by cell-lines personnel. For control purposes, the cell-lines department attempts to feed just enough alumina to last to the next scheduled crust break, and supervisors want each cell to "light" at least once a day.

Each 3-person stud-pulling crew works one 150-cell potline each day. One or two studs are removed from each anode and replaced each day in a regular pattern. Studs that are deep in the anode must be pulled before the anode is consumed to the level where the stud is exposed to the molten bath.

Stud pulling consists of three operations: (1) loosening the clamp that holds the stud to the anode buss; (2) positioning a special crane, seizing the stud with the crane wrench, rotating and pulling the stud, and placing it in a rack to be reconditioned; and (3) picking up a freshly cleaned stud with the crane wrench, positioning it in the same hole, but a higher level, and then clamping it to the bus. One stud puller works ahead of the crane and loosens the clamps on the studs to be removed. A second stud puller operates the crane. The third stud puller climbs onto the cell to fasten the clamp after the fresh stud is positioned. All three workers are exposed to air contaminants arising from the hot top of the anode. Also, as the hot stud is pulled through the semiliquid anode paste on top of the anode, considerable pollutants are generated. The crane cab is cooled with an evaporative cooler, but its positioning is such that the anode fumes are frequently drawn into the cab. All three members of the stud pulling crew are currently required to wear a half-mask respirator with dust-mist filters.

The briquetter drives a special truck through the potlines, adding briquettes of anode paste to the top of each anode daily. Briquette additions are coordinated with stud pulling, so that there is sufficient fresh, semiliquid anode paste on top of the cell to flow down into the hole left when the stud is pulled, filling the voids in the anode. The briquetter drives along the potline, charging briquettes onto the top of the cell with a special power conveyor from the truck. The briquetter then walks back along the cells distributing the briquettes with a long-handled rake. The briquetter works primarily at floor level but occasionally may climb onto the anode.

Each cell is tapped every other day to remove part (about 4000 lb) of the aluminum formed. On each shift the tapping crew taps one-third of the scheduled cells per day. The tapping crew places a large transfer crucible, one per cell, on the floor beside each cell. They break a hole in the crust, using an iron bar, and skim any impurities away. A long-spouted tapping crucible is positioned with the spout through the hole in the crust and penetrating into the metal in the bottom of the cell. The metal is sucked up the spout into the tapping crucible, then poured into the waiting transfer crucible. Two of the tappers are actually occupied tapping metal from the cells. The third tapper drives a large fork truck carrying the transfer crucibles to the plant foundry, where they are sampled and poured into a large holding furnace.

The cells in this plant are equipped with a typical VSS ventilation system to collect many of the contaminants that are released as part of the reduction process. This system consists of a flared skirt that surrounds the metal casing of the anode just above the molten bath (see Figure 7.1). This skirt nearly touches the frozen crust and is kept sealed by the cell operator by sweeping ore up against it. Gases and particulates captured under the skirt are drawn to a combustion chamber, where air is introduced and some volatile hydrocarbons and carbon monoxide are burned incompletely. The resulting combustion gases and other contaminants from the skirt system are carried away to the primary pollution control system, where they are cleaned before being exhausted to the atmosphere.

Nature of the Problem

The VSS plant that is the subject of this study has been in operation about 15 yr. The recognized health hazards, such as gaseous and particulate fluorides, heat stress, and noise, have been identified, evaluated, and largely controlled. These will not be addressed in this case study. However, the question of polynuclear aromatic hydrocarbon (PNA) exposures has not been examined. Because of the health concern about PNA exposures in potrooms, the management is very anxious to find out if a problem exists in this plant and, if so, how great it is. There is much confusion within the industry and among regulatory agencies about PNA exposures and how to measure and evaluate them.

Some state proposals are for application of the OSHA permissible exposure limit (PEL) for coal tar pitch volatiles (29 CFR 1910.1000, Table Z-1) to PNA exposures, since coal tar pitch is used as a binder in both anode and cathode paste. This standard requires the measurement of the benzene-soluble fraction of the total particulate matter collected on a filter and terms this fraction "coal tar pitch volatiles." Therefore, the analytical technique, and the standard itself, are nonspecific, and do not directly indicate the possible health hazard from the benzene-soluble particulate matter. Indeed, the documentation of the TLV for coal-tar pitch volatiles does not indicate if the benzene-soluble fraction is a good indicator of relative hazard or health effect. Moreover, much of the data on which the OSHA PEL was based was collected using high-volume air samplers and large filters.

Attempts to apply the analytical techniques used for large samples to personnel samples has generated conflicting results. Survey results have been presented showing that jobs with low potential for exposure to coal tar pitch volatiles, such as that of tappers, had exposures as great or greater than jobs with high exposure potential. For details, see tables and discussion in Shuler and Bierbaum.

This conflicting information has industry associations and management in various companies wondering if a problem really exists. At the same time, the management of this plant realizes that if a problem does exist, many radical changes in operations may be necessary. Good supporting exposure data will be required in order to gain corporate support and union cooperation to implement any needed changes. In addition, there are long-range plans being developed; these include optional process modifications that may affect PNA exposures. Options thought to reduce PNA exposures are included, and a complete current exposure profile will help evaluate the possible benefit of these options, once they are put in place.

Recently, NIOSH has developed improved analytical techniques for benzene-soluble particulate and used them in the study of coke oven exposures. These techniques still do not characterize the various PNAs in the benzene-soluble fraction, but the results are much more reproducible and realistic for personnel samples than those generated by the earlier procedure. At this time analytical procedures for specific PNAs on personnel samples are still being developed.

A preliminary evaluation of exposures to coal tar pitch volatiles in the potlines of this reduction plant has been performed (see Table 7.1). Based on this evaluation, some precautions have been instituted, including the use of half-mask respirators with dust-mist filters by those workers who must be above the cells where the volatiles from the tops of the anodes are most pervasive. This preliminary data has generated concern in management, particularly from the standpoint of the impact protective measures may have on future production.

TABLE 7.1
Preliminary Measurement of Exposures to Total Particulate and Benzene-Soluble Particulate

Job Title	Total Particulate (mg/m^3)	Benzene-Soluble Particulate (mg/m^3)
Cell operator	1.2	0.4
	3.5	0.2
	1.4	0.2
	1.9	0.3
Tapper	3.2	0.5
	1.1	0.1
	1.1	0.1
Briquette truck driver	1.1	0.2
	2.7	0.2
Stud puller	12.5	2.9
	5.1	0.8
	8.6	2.2
	4.3	0.8
	6.7	1.5
	5.6	1.0

Some precautions have been instituted as mentioned but, generally, the management is uncertain about the extent of the problem and the range of exposures within each job category.

Your job as industrial hygienist is to design and conduct a study to adequately characterize the coal tar pitch volatiles exposures of key groups of employees who work in the potlines. Your data will serve as the basis for an improved industrial hygiene program, including protective measures, employee education, and medical surveillance, where necessary. In addition, the results of your study will be used to plan future sampling and analysis aimed at identifying the most prevalent polynuclear aromatic hydrocarbons present in the coal tar pitch volatile portion of the exposure. This will allow more definitive answers regarding potential carcinogen exposure.

In your study, you should attempt to develop information that is statistically meaningful, but at the same time limit the overall completion time, since the management is anxious to include this information in their long-range planning. Your study results should also include short- and long-range recommendations for the industrial hygiene program, as well as suggestions for continued exposure monitoring.

Questions

1. What is your preliminary estimate of the degree of hazard from PNAs, coal tar pitch volatiles, and total particulates?
2. What immediate protective measures would you recommend, change, or add?
3. Prepare a justification for an in-depth exposure study.
4. How can you use the information in Table 7.1 to design a statistically meaningful study?
5. Describe the study plan including analytical methodologies, statistical sampling techniques and strategies,

types of data to be collected, and how they will be used. Particular attention should be placed on selecting statistically meaningful personal samples.

7.2. EVALUATION RESULTS

Preliminary Findings

Polynuclear aromatic hydrocarbons as a class of organic compounds include a number of potent carcinogens. PNAs are a major constituent of the incomplete combustion of organic matter, of cigarette smoke, and coal tar pitch. Because of this there has been considerable concern about exposure to PNAs in aluminum reduction plants, especially in the reduction cell buildings, the "potlines." The standard currently applied to evaluate PNA exposures by OSHA and state regulatory agencies is the OSHA PEL for coal tar pitch volatiles.

The ACGIH has identified coal tar pitch volatiles as recognized human carcinogens with a TLV of 0.2 mg/m^3. There is not an OSHA PEL for PNAs, except for the coal tar pitch volatiles standard. OSHA has proposed a PEL for one PNA, benzo[a]pyrene, of 0.2 $\mu g/m^3$. Beyond these, there is little guidance on exposure limits, although the National Academy of Sciences has published a monograph on particulate polycyclic organic matter, which identifies a number of potent carcinogens among the PNAs.

After a walk-through survey, the initial exposure data (Table 7.1) was reviewed in order to determine if any immediate measures were required to better protect the employees while a more extensive evaluation of exposures is carried out. Since environmental data frequently fits a log-normal distribution, the benzene-soluble particulate matter data was grouped and plotted using the techniques in Appendix I of Leidel et al. Figure 7.3 shows the estimated best fit lines for stud pullers (up-

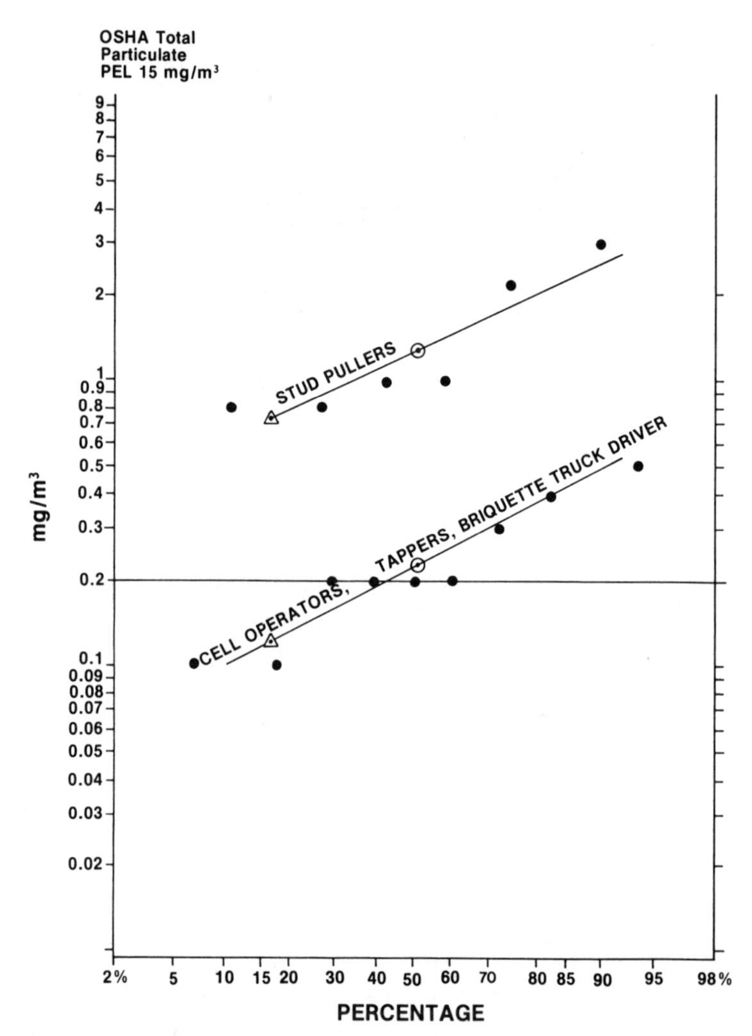

FIGURE 7.3. Cumlulative percent frequency plot of preliminary benzene-soluble particulate exposure data for stud pullers, and for cell operators, tappers, and briquette truck driver.

per line) and cell operators, tappers, and briquette truck driver (lower line). Since the data appeared to fit a log-normal frequency plot and no apparent tailing was present at the lower end, it was assumed that a two-parameter log-normal distribution would be adequate for this preliminary analysis. The geometric means (GM) and geometric standard deviations (GSD) for the two exposed groups were read directly from the graph (see Table 7.2).

Protective measures to prevent harmful effects from exposure to health hazards should be designed to protect against the periodic high exposures that occur during the normal course of work operations. In other words, the time-weighted average exposures, which are measured during normal operations, are distributed, in some way, from low to high; protective measures should be directed to reducing those highest TWAs to safe levels. If some information is available about the shape of the distribution, it is possible to state a value that would be expected to contain a stated proportion of the population of TWA values. This value is termed a tolerance limit, and it is possible to specify both upper and lower tolerance limits. For purposes here, the upper tolerance limit is sufficient.

TABLE 7.2
Benzene-Soluble Particulate, Preliminary Data Analysis[a]

Exposure Group	Geometric Mean (mg/m^3)	Geometric Standard Deviation	Upper Tolerance Limit
Cell operators, tappers, briquette truck driver	0.21	1.7	1.1
Stud pullers	1.37	1.7	10.1
OSHA PEL for coal tar volatiles	0.2		

[a]Source: Natrella, M.G., 1963.

For example, if the population of TWA values followed a normal distribution, and the "population" mean and standard deviation were known, then 95% of the TWAs would be less than the mean plus 1.64 times the standard deviation. This would be analogous to a 95% upper tolerance limit on that population.

Since only estimates for the mean and standard deviation (i.e., from a sample) for the data in Table 7.1 can be derived, certainty about the value bounding a particular fraction of the population of TWA values decreases. Therefore, it is necessary to increase the multiplier from the familiar z (i.e., 1.64) statistic to a *K* factor or tolerance factor, which takes into account the size of the sample used to estimate the mean and standard deviation. Both Leidel and Busch, and Natrella give good discussions of the basis for and use of tolerance limits. An abbreviated table (Table 7.3) of *K* factors for one-sided tolerance limits, taken from Natrella, is included for illustration. For small sample sizes, the extremes of any distribution are very uncertain, so for this preliminary evaluation, the 95% upper tolerance limit for the benzene-soluble particulate-matter TWAs was calculated as a basis for evaluating the existing protective measures in the plant.

These values are based on small numbers of samples taken over a short span of time, so they must be used with discretion. In this case, these data suggest the upper range of exposures that may be encountered by workers in those groups and indicate the level of respiratory protection that might be required. On this basis, it was recommended that the stud pullers be provided with full-face respirators equipped with combination high efficiency particulate arrester (HEPA) filter - organic vapor chemical sorbent cartridges. Organic vapor cartridges were specified to remove PNAs that may volatilize after being filtered as particulates. Full-face respirators are assigned a protection factor of 50 and therefore would be adequate for

TABLE 7.3
***K* Factors for One-sided Tolerance Limits for Normal Distributions**[a]

	P for $\gamma = 0.95$				
n	0.75	0.90	0.95	0.99	0.999
3	3.804	6.158	7.655	10.552	13.857
4	2.619	4.163	5.145	7.042	9.215
5	2.149	3.407	4.202	5.741	7.501
6	1.895	3.006	3.707	5.062	6.612
7	1.732	2.755	3.399	4.641	6.061
8	1.617	2.582	3.188	4.353	5.686
9	1.532	2.454	3.031	4.143	5.414
10	1.465	2.355	2.911	3.981	5.203
11	1.411	2.275	2.815	3.852	5.036
12	1.366	2.210	2.736	3.747	4.900
13	1.329	2.155	2.670	3.659	4.787
14	1.296	2.108	2.614	3.585	4.690
15	1.268	2.068	2.566	3.520	4.607
16	1.242	2.032	2.523	3.463	4.534
17	1.220	2.001	2.486	3.415	4.471
18	1.200	1.974	2.453	3.370	4.415
19	1.183	1.949	2.423	3.331	4.364
20	1.167	1.926	2.396	3.295	4.319
21	1.152	1.905	2.371	3.262	4.276
22	1.138	1.887	2.350	3.233	4.238
23	1.126	1.869	2.329	3.206	4.204
24	1.114	1.853	2.309	3.181	4.171
25	1.103	1.838	2.292	3.158	4.143
30	1.059	1.778	2.220	3.064	4.022
35	1.025	1.732	2.166	2.994	3.394
40	0.999	1.697	2.126	2.941	3.866
45	0.978	1.669	2.092	2.897	3.811
50	0.961	1.646	2.065	2.863	3.766

[a]Where P is the proportion of population values to be included, γ is the probability that the tolerance limit includes the stated proportion, and n is sample size. Adapted from Natrella, 1963.

exposures up to the 95% upper tolerance limit of 10.1 mg/m^3 (50 $\times$ the OSHA PEL for coal tar pitch volatiles).

It was also recommended that cell operators, tappers, and briquette truck drivers be offered, but not required to wear, half-mask respirators equipped with combination HEPA filter–organic vapor chemical sorbent cartridges. Half-mask respirators are assigned a protection factor of 10 and therefore would be adequate for exposures greater than 95% of those expected. The data indicate that respirators should be required, but, after frank discussions with management, it was agreed that support for a full-blown respirator program throughout the entire potroom work area would be better after a more complete assessment of exposures. The management organization argued that employee cooperation and support from first-line supervision would be better if the facts to support such a program were gathered. (It is worth noting that the industrial hygiene staff, at the plant and at the corporate level, agreed with this approach. Both groups were prepared to make employees aware of the data and conclusions to date as part of training for recommended respirator use.)

It was also recommended that the engineering department begin preliminary investigations into methods for providing air-supplied respirators for the stud-pulling crews, since they perform hot, heavy work, and a full-face negative-pressure respirator is bound to add significantly to the stress of the job.

Other initial recommendations were that (1) all potroom personnel be given respirator training to acquaint them with the proper use of the appropriate respirators and the reason for respiratory protection, as required by OSHA regulations; (2) the importance of good personal hygiene be stressed in the respirator training program, as well as during initial orientation of all new potroom personnel, since exposure to coal tar pitch products is known to cause skin irritation and photosensitivity.

In-Depth Studies

A number of possible goals were put forward for the in-depth sampling program. First, a reliable estimate of exposures in each job within the potroom was needed, in order to accurately assess the degree of hazard, and plan effective control measures. Second, reliable exposure data would provide the basis for "selling" difficult protective measures to management and employees. Third, if significant sources of contaminants could be identified, control efforts could be directed to those first. Finally, preliminary planning for previously mentioned process modifications, which could favorably affect potroom exposures to PNAs, was underway; if reliable estimates of "before" and "after" exposures were available, they would help, in the future, to establish whether the reduced exposures markedly decrease health risks.

Based on the process, and the limited information in Table 7.1, several factors appear to affect the level of exposure to polynuclear aromatic hydrocarbons. First, the nature of the work carried out in each job category, as well as any individual approaches to the job duties would certainly affect the exposure level. Second, the status of the aluminum reduction production operation would determine the amount of volatile organic material released. For example, if the tops of the anodes were hotter, more volatiles would be driven off before being incorporated into the matrix of the anode, or if the hood or skirt system were not kept sealed, more contaminants, including PNAs, would escape into the work room atmosphere instead of being captured in the pollution control system. Third, if the PNAs were condensed and adsorbed onto the surface of other particulate matter, work operations that stirred up settled dust could increase exposures. Finally, ambient weather conditions, particularly temperature, could affect potroom temperatures, anode temperatures, and even ventilation air flow, since, in hot weather there is a tendency to leave large doors and windows open.

The following objectives and statistical sampling procedures were established for the first year of the exposure distribution monitoring program:

1. Collect sufficient samples to assess the exposures to coal tar pitch volatiles in each potroom job and establish a 95% upper tolerance limit for "hot weather" and "cold weather."
   a. These tolerance limits would be established in the same manner as the estimates made in evaluating the initial exposure data from 7.1, except that the geometric mean and geometric standard deviation would be calculated by making a log transform of the individual 8-hr TWA exposure results, and the mean and standard deviation of the logs would then be calculated.
   b. The actual number of samples from each job category was determined by examining Table 7.3. These factors decrease as sample size increases, thereby giving tighter estimates. Available resources and analytical costs were considered, and the target selected was six 8-hr TWAs for each job category for each season studied. Of course, the larger the sample size the smaller the K or multiplier, which means a potentially smaller value for the upper 95% tolerance limit. This could mean smaller control costs; however, it definitely means greater sampling costs. Note that as sample size increases, K approaches the normal z statistic.
   c. In an actual production operation, it is very difficult to control the variables that may affect exposure.

Consequently, the approach taken was to stratify the sampling for each job category.

(1) Each job category (cell operator, stud puller, tapper, briquette truck driver) represented one stratum.

(2) Within each stratum, individuals were randomly chosen by selecting sampling days at random.

d. The objective was to avoid sampling the same individual multiple times. Since the plant operates a rotating shift, where the entire crew works a week on each shift and is off the fourth week, this tended to prevent sampling the same person repeatedly.

(1) Where the work group was large enough, for example, cell operators, the work location was selected randomly as well.

(2) Where the work group was too small to select individuals randomly, for example, tappers ($n = 3$), random days were relied upon solely to avoid bias in selection.

2. Collect data to allow analysis of correlations between process variables and exposure levels for each job. Although process variable data are not presented here in order to reduce the complexity of the case study, the following data were recorded during the study:

a. Production rates.

b. Cell amperage and voltage.

c. Frequency of "lights."

d. Individual work rates, such as cells tapped, studs pulled, cells briquetted, and so on.

3. Finally, since weather effects were expected, both summer and winter surveys were planned. Weather data was

obtained from the weather-monitoring equipment operated for the ambient air sampling program. Cold-weather data was collected in March, hot-weather data in August.

Personnel exposures to total particulates and benzene-soluble particulate matter were measured using personnel sampling pumps, sampling at 2 L/min. Following the recommended NIOSH procedures, samples were collected on 37-mm diameter glass-fiber filters, backed with silver-membrane filters in 37-mm polystyrene plastic filter holders.

After the filters were weighed to determine total particulate, benzene-soluble particulate matter was determined. The technique used was that recommended in the NIOSH *Criteria for a Recommended Standard: Occupational Exposure to Coal Tar Products* (1978). The results of the exposure monitoring are given in Tables 7.4 and 7.5.

Questions

1. Discuss the decision for respirator use that was based on statistical analysis of the preliminary results. Would you have tried to convince management of a different program? What would it be?
2. How will you analyze the data in Tables 7.4 and 7.5? Discuss the treatment you will give each type of data. Some data is not presented in this case study, for example, operational and process variables (objective 2 above). However, explain what you would do with the data if it were present.
3. What conclusions can be drawn about 8-hr TWA exposures for each job category? What statistical methods will you use to compare process, weather, and other exposure variables with exposure levels?

TABLE 7.4
Total Particulate (mg/m^3) and Benzene-Soluble Particulate (mg/m^3) from In-Depth Studies, Cold Weather[a]

Job	Date	Crew	Total TWA	Benzene TWA
Cell operator	3/1	Section 1	9.54	0.67
	3/1		7.33	0.36
	3/4	Section 2	2.95	0.19
	3/4		2.91	0.15
	3/7	Section 1	2.81	<0.01
	3/7		7.94	0.16
	3/9	Section 2	2.63	0.44
	3/9		6.11	0.01
	3/16	Section 1	3.41	0.14
	3/16		0.77	0.03
	3/17	Section 2	6.50	0.35
	3/17		2.52	0.03
Stud puller	3/1	Line A	6.53	2.16
	3/1		11.53	4.55
	3/4		7.55	2.38
	3/4		4.41	0.33
	3/7		5.34	1.54
	3/7		5.80	0.68
	3/9	Line B	6.86	0.56
	3/9		4.86	1.57
	3/16		8.37	0.55
	3/16		7.13	1.95
	3/18		7.90	0.45
	3/18		7.98	0.63

(Continued)

TABLE 7.4 (Continued)

Job	Date	Crew	Total TWA	Benzene TWA
Briquetter	3/1		2.63	0.37
	3/3		1.15	0.11
	3/4		0.79	0.19
	3/7		7.14	0.23
	3/9		3.57	0.18
	3/11		0.90	0.24
	3/16		5.64	0.02
	3/18		0.35	0.31
	3/19		3.68	0.22
Tapper	3/9	Tapping crew	9.33	0.06
Tapper	3/9		2.95	0.04
Truck driver	3/9		1.04	0.04
Tapper	3/10		2.08	0.26
Tapper	3/10		2.23	0.03
Truck driver	3/10		2.26	0.15
Tapper	3/12		5.14	0.59
Tapper	3/12		12.27	0.10
Truck driver	3/12		1.93	0.05
Tapper	4/2		4.99	0.23
Tapper	4/2		0.77	0.19
Truck driver	4/2		1.10	0.01
Tapper	4/3		1.41	0.15
Tapper	4/3		1.99	0.01
Truck driver	4/3		0.92	0.14
Tapper	4/4		8.54	0.16
Tapper	4/4		9.20	0.16
Truck driver	4/4		2.49	0.05

[a]TWA averages are for sampling periods ranging from 6 to 8 hr.

TABLE 7.5
Total Particulate (mg/m³) and Benzene-Soluble Fraction (mg/m³) from In-Depth Studies, Warm Weather[a]

Job	Date	Crew	Total TWA	Benzene TWA
Cell operator	7/26	Section 1	3.15	0.18
	7/26		1.44	0.55
	7/28	Section 2	1.18	0.36
	7/28		0.62	0.18
	8/02	Section 1	5.55	0.05
	8/02		6.59	0.13
	8/04	Section 2	3.79	0.22
	8/04		3.33	0.01
	8/09	Section 1	1.98	0.22
	8/09		7.89	0.25
	8/11	Section 2	1.78	0.21
	8/11		1.95	0.25
Stud puller	7/27	Line A	8.54	0.66
	7/27		23.52	8.49
	7/27	Line B	2.84	0.12
	7/27		6.78	2.28
	8/05	Line A	4.34	0.37
	8/05		3.22	0.52
	8/05	Line B	2.93	0.23
	8/05		6.74	1.34
	8/10	Line A	7.75	1.54
	8/10		5.64	0.87
	8/10	Line B	8.74	1.82
	8/10		8.27	1.03
Briquetter	7/26		2.14	0.21
	7/27		0.96	0.24
	7/28		0.42	0.23
	8/02		2.56	0.20
	8/04		1.31	0.19
	8/09		0.75	0.17
	8/10		0.92	0.21
	8/11		1.25	0.19

(Continued)

TABLE 7.5 (Continued)

Job	Date	Crew	Total TWA	Benzene TWA
Tapper	8/16	Tapping	4.55	0.15
Tapper	8/16		1.25	0.01
Truck driver	8/16		2.96	0.14
Tapper	8/18		1.72	0.19
Tapper	8/18		2.16	0.23
Truck driver	8/18		1.05	0.15
Tapper	8/23		1.83	0.57
Tapper	8/23		5.38	0.10
Truck driver	8/23		1.58	0.01
Tapper	8/25		2.02	0.03
Tapper	8/25		1.77	0.26
Truck driver	8/25		1.34	0.07
Tapper	8/30		4.93	0.04
Tapper	8/30		10.67	0.01
Truck driver	8/30		1.87	0.04
Tapper	9/01		6.91	0.13
Tapper	9/01		4.42	0.06
Truck driver	9/01		0.73	0.11

[a]All TWA are for sampling periods ranging from 6 to 8 hr.

4. What additional protective measures would you recommend based on the in-depth study results?
5. Should personnel in any of the job categories be provided additional protection?
6. Can future benzene-soluble particulate exposure levels be estimated from total particulate exposures?
7. What additional industrial hygiene controls could be used?

7.3. CONTROL STRATEGIES

Discussion

The log-normal distribution was used to describe the in-depth data. In order to calculate the geometric mean and geometric standard deviation, the transformation $y = \log x$ was applied to the data, and the mean and standard deviation of y calculated. The geometric mean of x is antilog $\bar{y}$, and the geometric standard deviation of x is antilog s_y. The resulting parameters can be plotted on log-probability paper to illustrate the distribution of the data. The geometric mean (GM) is plotted at 50%, and the GM plus and minus one geometric standard deviation (GSD) are plotted at 15.9 and 84.1%, respectively. A straight line connecting these three points gives the cumulative log-normal distribution representing the data. Cumulative log-normal plots of the data for each job category are shown in Figures 7.4, 7.5, 7.6, and 7.7.

The 75 and 95% upper tolerance limits in Table 7.6 are particularly useful as indicators of the proportion of the population of exposures that are expected to exceed the appropriate standards. These are calculated as discussed in Section 7.2 (mean + K times standard deviation). They show that coal tar pitch volatiles, as measured by the benzene-soluble particulate matter, pose a serious exposure for all job categories, especially the stud pullers. At the same time, total particulate exposures for all job categories, except the briquette and tapping truck drivers, would be expected to exceed the OSHA PEL and ACGIH TLV more than 5% of the time. Notice that if the upper tolerance limits are plotted on the cumulative frequency plots, they lie above the line representing the log-normal distribution. This is due to the uncertainty about the upper tolerance limit because of the small sample size.

The plotted distribution, and the calculated tolerance limits, indicate an "unacceptable" (Leidel and Busch) exposure

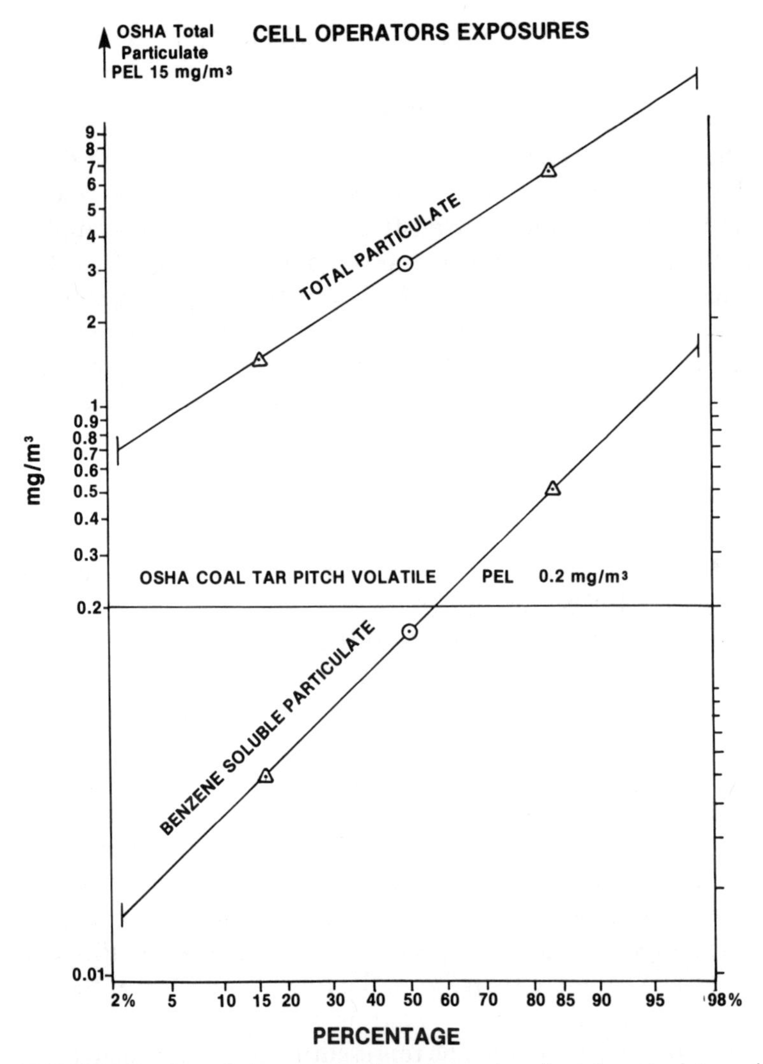

FIGURE 7.4. Cumulative percent frequency plot of total particulate and benzene-soluble particulate exposure data for cell operators.

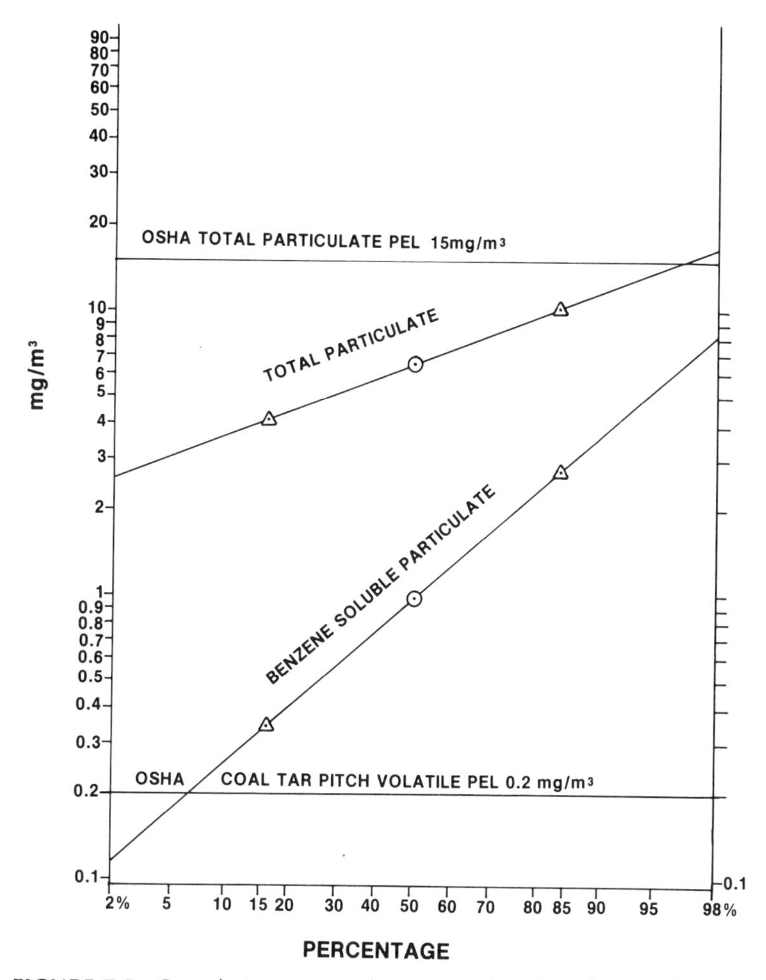

FIGURE 7.5. Cumulative percent frequency plot of total particulate and benzene-soluble particulate exposure data for stud pullers.

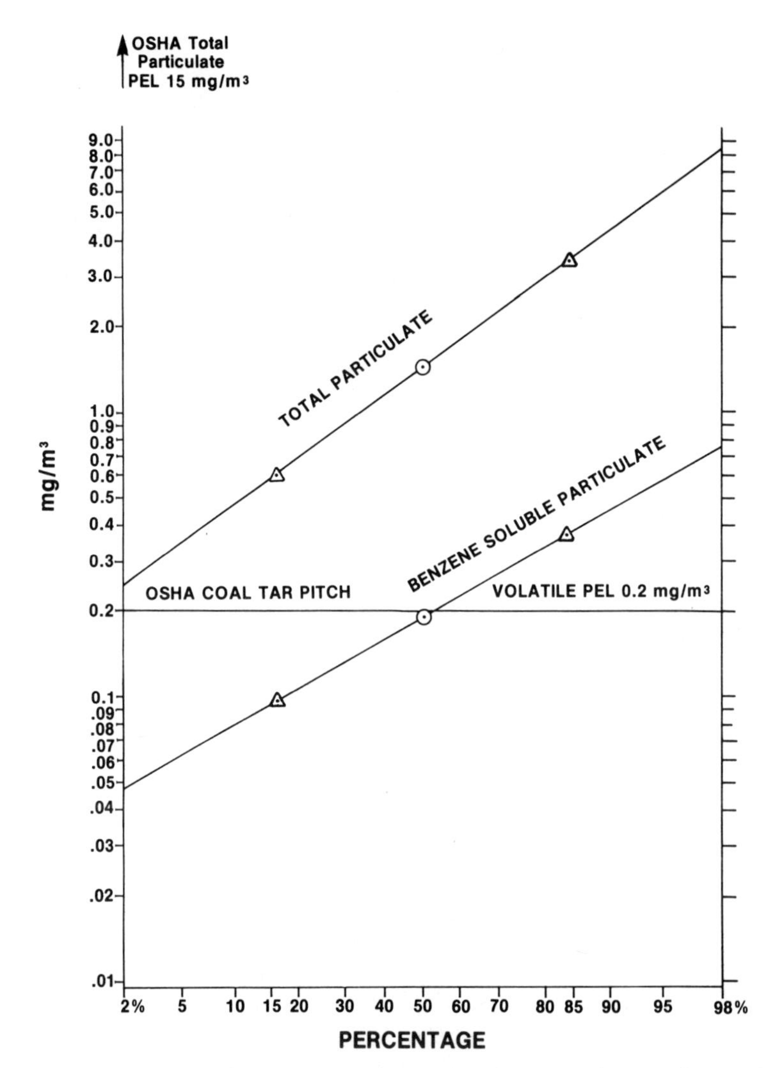

FIGURE 7.6. Cumulative percent frequency plot of total particulate and benzene-soluble particulate exposure data for briquette truck drivers.

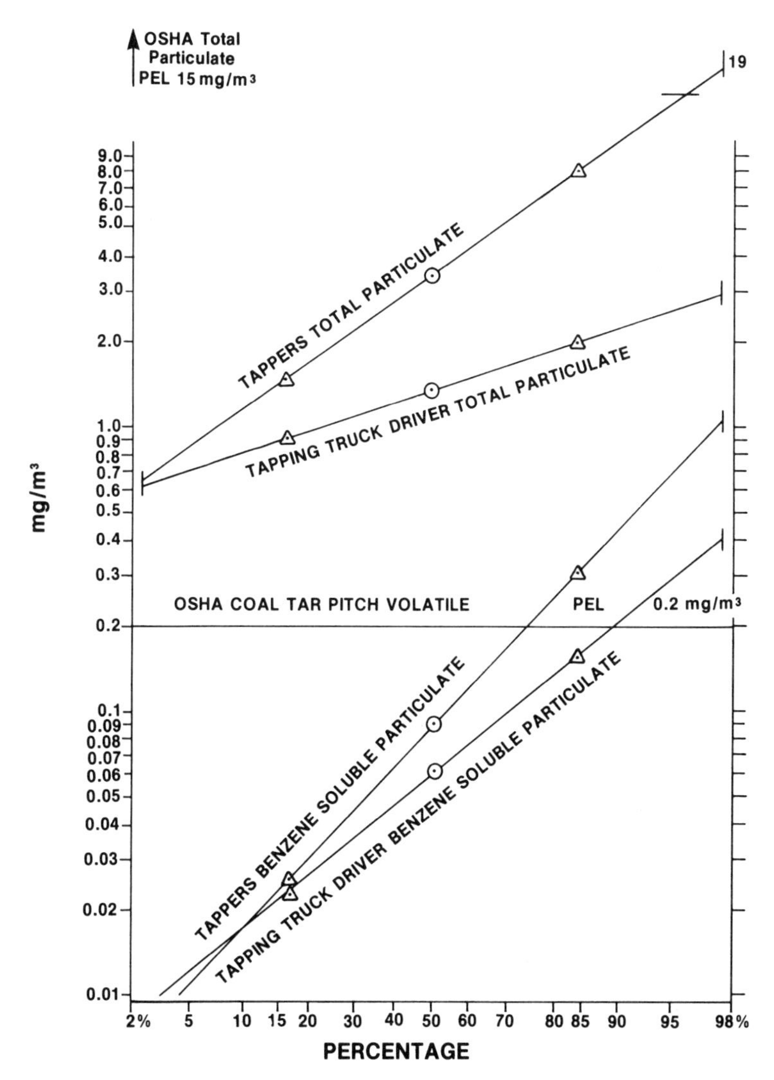

FIGURE 7.7. Cumulative percent frequency plot of total particulate and benzene-soluble particulate exposure data for tappers and tapping truck drivers.

TABLE 7.6
Average 8-Hr TWA Exposures[a]

	Total Particulate				Benzene Soluble			
Job Category	GM	GSD	75%	95%	GM	GSD	75%	95%
Cell operator	3.13	2.09	7.13	17.27	0.15	3.18	0.53	2.11
Stud pullers	6.50	1.57	10.75	18.47	0.97	2.63	2.85	9.07
Briquette truck driver	1.50	2.37	4.30	8.44	0.18	1.87	0.39	0.64
Tapping crew								
Tappers	3.45	2.16	8.13	20.36	0.09	3.25	0.35	1.41
Tapping truck driver	1.47	1.55	2.68	4.90	0.06	2.63	0.22	0.81
OSHA PEL		15.0				0.2		
ACGIH TLV		10.0				0.2		

[a]All values are in mg/m^3, except GSD.

distribution. At least 25% of the TWA exposures to benzene-soluble particulate in all job categories exceed the OSHA PEL. In the case of total particulate matter, the tappers and stud pullers definitely have an unacceptable exposure. Cell operators may fall into Leidel and Busch's marginal exposure category for total particulate. At a minimum, future sampling should be designed to clarify sources of total particulate exposure so that the exposure can be reduced.

As stated previously, the operational data was collected and analyzed but not presented in this case study. During the first year's survey, most operational data did not vary enough to allow meaningful correlations with exposures. Stud-puller exposures did correlate positively with work load. A regression of benzene-soluble particulate matter versus number of studs pulled gave an $r = 0.95$. Work load versus exposure level data did not correlate strongly for other job categories.

However, operational data should always be considered. Indeed, historical data on pollution emissions from years previous to the study did show that "upset" operations, such as high anode temperatures, increase the levels of PNAs in the potroom atmosphere.

There was a poor correlation between total particulate versus benzene-soluble particulate exposure, indicating that future monitoring for exposure to benzene-soluble particulate should be continued, rather than relying on a total particulate exposure value to indicate relative exposure. This was unfortunate, since the benzene-soluble particulate analysis technique is more difficult to accomplish. Consequently, this is an important result.

The data did not show a significant difference between exposure levels during cold and hot weather. It was concluded that most of an exposure is due to the work carried out by the employee (i.e., local conditions), rather than ambient concentrations in the work room. Operational data directly related to work load, such as number of studs pulled, was correlated with exposure (as noted before). For future monitoring, data collection should primarily focus on individual work loads, with only one or two longer-range plant operational parameters, such as the current monthly production level or current monthly energy consumption gathered for correlation studies.

Based on the data in Table 7.6, the following recommendations were presented to management:

Control Measures

1. All cell lines' employees working in the cell lines should be required to wear half-mask respirators with HEPA filter - organic vapor chemical sorbent cartridges while working in the cell lines. The stud-pulling crew should

be provided air-supplied or powered air-purifying respirators. Providing air lines to stud-pulling crews was not possible in this plant due to the cell layout and electrical hazards, so full-face powered air-purifying respirators with HEPA filters were selected. A complete respirator program should be developed.

2. Clean break rooms with hand-washing sinks and air-conditioned filtered air should be provided for breaks and eating and smoking.
3. Consumption of food and drink should not be permitted in the cell rooms.
4. All employees should be given an annual orientation to possible health hazards of exposure to polynuclear aromatic hydrocarbons as found in benzene-soluble particulate matter. Special emphasis should be given to the need for good personal hygiene, the importance of cleaning the skin well, and the use of respiratory protection.
5. A medical-monitoring program should be developed and instituted to monitor cell lines employees for possible effects of exposure to polynuclear aromatic hydrocarbons. This should include preplacement and annual history and examination, with special emphasis on the respiratory system and skin. A physician should be retained to supervise the nurses and direct the medical-monitoring program. All medical records should be placed under the physician's direct control. These last two points are raised because an effective medical-monitoring program depends on maintaining confidentiality of medical records. When records are accessible to other, nonmedical employees they are open to abuse, and employees become reluctant to cooperate. Also, nurses under a physician's supervision are in a better position to exercise good professional judgment, free from management pressures, and have some protection from liability.

6. A program should be undertaken to characterize the nature of the polynuclear aromatic hydrocarbons in the benzene-soluble particulate matter more completely.

Long-Term Solution

As a result of this study, these recommendations were adopted. In addition, a decision was made to proceed with a major process change that would potentially lower the levels of benzene-soluble particulate matter in the potrooms.

Alternative Solutions

Other potential solutions to the polynuclear aromatic hydrocarbon exposures considered included:

1. Change to a prebaked cell, with state-of-the-art controls in the anode plant.
2. Place hoods above existing anode tops to capture PNAs as they are released.
3. Develop robots for stud-pulling and similar operations to remove the employee from the exposure.

While alternative 3 was strongly considered, the process change selected was carried out first, as this would permit further modification or plant operations more easily. Options 1 and 2 were rejected as impossible in an existing plant without virtual rebuilding, and option 2 would require stud pullers to either remove or enter the hood in order to work, limiting the needed exposure reduction.

Additional Remarks

An additional refinement to deciding the number of samples in each job category would be to plan to collect enough samples to permit statistical tests of exposures before and after significant process changes.

The means for job categories would be compared using a t test on the log-transformed data. The null hypothesis is H_0: $\mu_1 = \mu_2$, and the alternative hypotensis H_a: $\mu_1 > \mu_2$. The chosen alpha, or type I error risk, might be set at 0.10, and the beta, or type II risk, at 0.05. This is because if the null hypothesis, that there is no difference between exposures after process changes, is erroneously accepted (type I error), it will eventually become apparent (with continued monitoring) that exposures have changed, and action will be taken. On the other hand, if it is incorrectly concluded that future exposures have been reduced, protective measures may be relaxed when still required. Therefore, type II error should be minimized.

Experimental design calculations using the estimated GM and GSD from Section 7.1 (see W. J. Diamond) indicate that, at the 95% confidence level, 11 total exposure samples for cell operators, tappers, and briquette truck drivers would be adequate to detect a difference of 1 × the PEL for coal tar pitch volatiles, while 17 total samples would be required for stud pullers. Only 8 stud puller samples would be required to detect a larger exposure difference of 1.5 × the PEL for coal tar pitch volatiles.

Unfortunately, the geometric standard deviation of the time-weighted average exposures to both total particulate and benzene-soluble particulate were much greater than estimated from the preliminary data. Recalculating the required sample sizes using the GSDs from the in-depth data indicates that from 24 to 72 samples will be required after any process change, to detect an average change in benzene-soluble particulate exposures of 0.2 mg/m^3. If the change is much greater than 0.2 mg/m^3, fewer samples would be required to confirm the reduction.

REFERENCES

American National Standards Institute. *ANSI 788.2 Recommended Practices for Respiratory Protection*, New York, 1980.

Burgess, W.A. *Recognition of Health Hazards in Industry*, Wiley, New York, 1981, pp. 143-147.

Clayton, G.D., and F.E. Clayton, eds. *Patty's Industrial Hygiene and Toxicology* 3rd ed., Vol 2C, Wiley, New York, 1982.

Committee on Biologic Effects of Atmospheric Pollutants. *Particulate Polycyclic Organic Matter*, National Academy of Sciences, Washington, DC, 1972.

Craver, J.S. *Graph Paper from Your Copier*, H.P. Books, Tucson, AZ, 1980.

Diamond, W.J. *Practical Experiment Designs for Engineers and Scientists*, Lifetime Learning Publications, Belmont, CA, 1981.

International Primary Aluminium Institute. *Health Protection in Primary Aluminium Production*, Vol 1, Proceedings of a Seminar in Copenhagen, June 28-30, 1977, and Vol. 2, Proceedings of a Seminar in Montreal, September 22-24, 1981.

International Primary Aluminium Institute Health Committee Review. *The Measurement of Employee Exposures in Aluminium Reduction Plants*, London, 1982.

Karsten, K.P. *Aluminum*, Chap. 8 in *Industrial Hygiene Aspects of Plant Operations*, Volume 1, *Process Flows*, L.V. Cralley and L.J. Cralley, Eds., Macmillan, New York, 1982, pp. 47-59.

Leidel, N.A., K.A. Busch, and J.R. Lynch, *Occupational Exposure Sampling Strategy Manual*, USDHEW, NIOSH, Publication No. 77-173, Cincinnati, OH, 1977.

Leidel, N.A. and K.A. Busch, "Statistical design and data analysis requirements," in L.V. Cralley and L.J. Cralley, Eds, *Patty's Industrial Hygiene and Toxicology*, Vol III A, 2nd ed., *Theory and Rational of Industrial Practice*, Wiley, New York, 1985.

National Institute of Occupational Safety and Health. *NIOSH Manual of Analytical Methods*, 3rd ed., P.M. Eller, Ed., USDHHS, CSC, NIOSH, Cincinnati, OH, 1984.

Natrella, M.G. *Experimental Statistics*, National Bureau of Standards Handbook 91, U.S. Government Printing Office, Washington, DC, 1963.

Rush, D., Russell, J.C., and Iverson, R.E. "Air pollution abatement on primary aluminum potlines: Effectiveness and cost," *J. Air Pollut. Control Assoc.*, 23:2, 1973.

Sheehy, J.W. *Occupational Health Control Technology for the Primary Aluminum Industry*, USDHEW, NIOSH, Publication No. 83-115, Cincinnati, OH, 1983.

Shuler, P.J., and Bierbaum, P.J. *Environmental Surveys of Aluminum Reduction Plants*, USDHEW, NIOSH, Publication No. 74-101, Cincinnati, OH, 1974.

United States Department of Health, Education, and Welfare. *Criteria for a Recommended Standard: Occupational Exposure to Coal Tar Products*, NIOSH, Publication No. 78-107, Cincinnati, OH, 1978.

United States Department of Health, Education, and Welfare. *The Industrial Environment: Its Evaluation and Control*, NIOSH, Cincinnati, OH, 1973.

INDEX